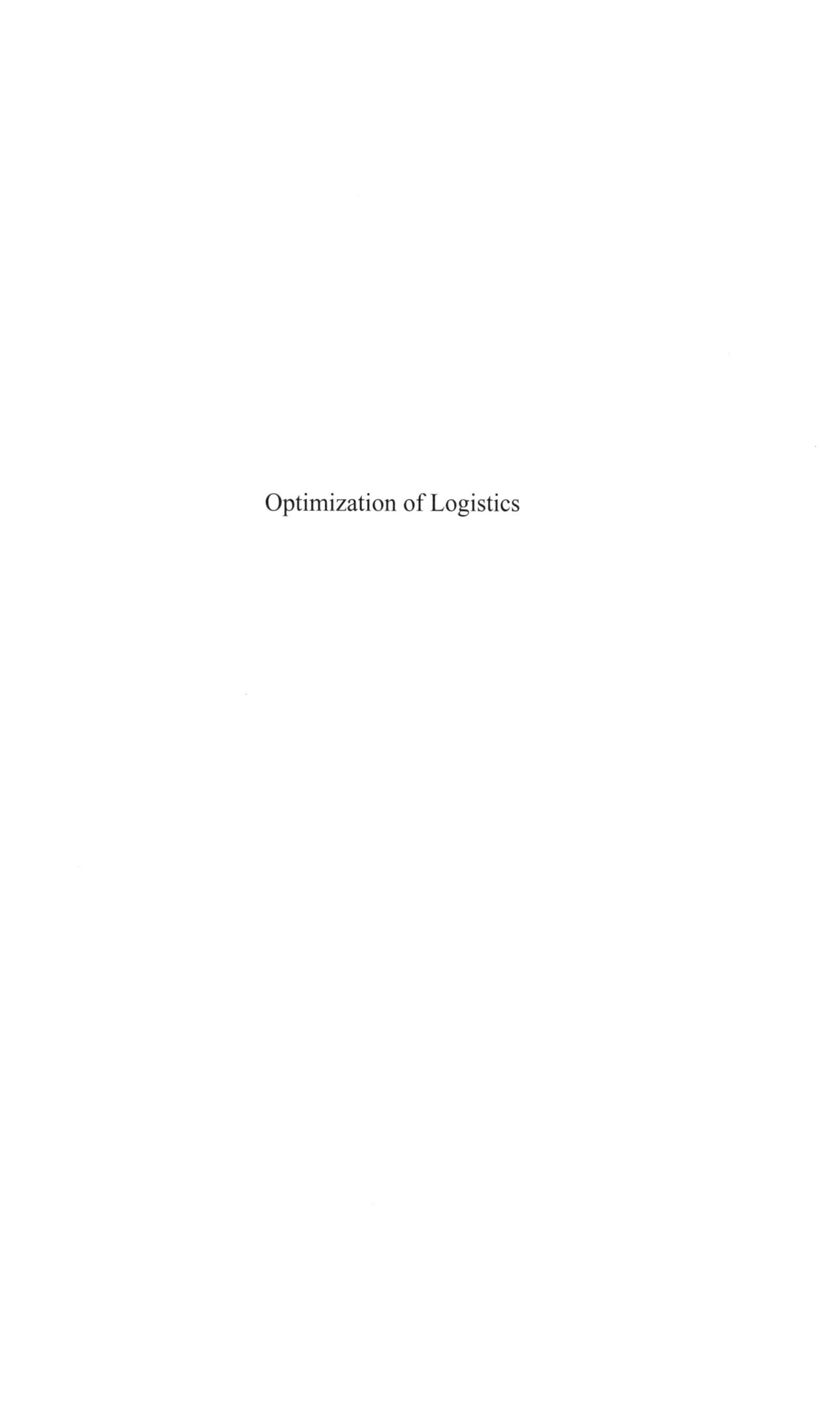

Optimization of Logistics

This book is dedicated to the memory of Mohamad Chehade.

It is also dedicated to our respective families, who supported and encouraged us:

– Sara, Lehna and the whole Yalaoui family;

– Farah, Lana and the whole Chehade family;

– Stéphanie, Laura, Aurélien and the whole Amodeo family.

We thank everyone for their participation in proofreading this work, and for their support and invaluable assistance: Frédéric Dugardin, Romain Watier, Stéphanie Dussolier, Julie Rubaszewski, Yassine Ouazene, Andrès Felipe Bernate Lara, Naïm Yalaoui, Farah Belmecheri, Atefeh Moghaddam and Slim Daoud.

Finally, we thank all our colleagues as well as our academic and industrial partners.

Optimization of Logistics

Alice Yalaoui
Hicham Chehade
Farouk Yalaoui
Lionel Amodeo

Series Editor
Jean-Paul Bourrières

First published 2012 in Great Britain and the United States by ISTE Ltd and John Wiley & Sons, Inc.

ISTE Ltd
27-37 St George's Road
London SW19 4EU
UK

www.iste.co.uk

John Wiley & Sons, Inc.
111 River Street
Hoboken, NJ 07030
USA

www.wiley.com

Library of Congress Control Number: 2012946446

British Library Cataloguing-in-Publication Data
A CIP record for this book is available from the British Library
ISBN 978-1-84821-424-8

Printed and bound in Great Britain by CPI Group (UK) Ltd., Croydon, Surrey CR0 4YY

Table of Contents

Introduction

To remain competitive, businesses must design, produce and distribute their products and services while taking account of delays and increasingly restrictive quality requirements. To succeed, they put in place new technologies and methods to continually improve their production and distribution tools. Such a system is called a production system, or more commonly a logistics system. These systems combine the set of processes due to which a business produces a product or service, thus allowing it to satisfy a demand.

The components of a logistics system can be divided into two broad categories, thereby defining a conceptual model [AMO 99]: the physical system and the production control system (Figure I.1).

The role of the physical system is to transform the raw materials and components into finished products. It is composed of human resources, machines, transport systems, storage systems and numerous other elements. The major decisions at the control system level may be classified into three categories: strategic, tactical and operational decisions [PAP 93]. Design and management problems (Figure I.2) are divided into three distinct but interacting stages: long-term, medium-term and short-term decisions.

For long-term decisions, a time frame of several years may be considered. Decisions are to be made regarding the initial design of the system as well as the functional modifications to be made. These decisions may be preliminary, with some data, or detailed decisions. Preliminary decisions are characterized by approximate data, which require the testing of a significant number of solutions in order to obtain the best design. For detailed decisions, this consists of evaluating the most promising solutions, as the data become more precise.

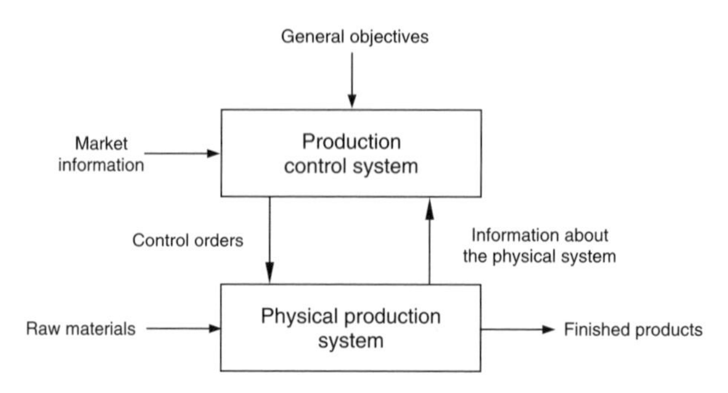

Figure I.1. *Conceptual model of a production system [AMO 99]*

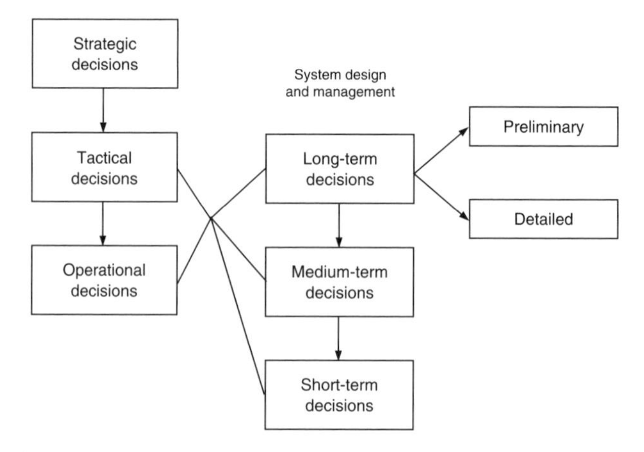

Figure I.2. *Position of design and management problems [PAP 93]*

Medium-term decisions are distributed across a time frame ranging from several weeks to several months. They essentially consist of determining the types of products and the planning of their production in certain time frames, taking into consideration several parameters such as production targets, constraints on resource capacity and costs.

Regarding the short-term decisions, the time frame varies from several hours to several days. These decisions concern the distribution of resources in the system, scheduling production tasks, reactions to unforeseen events, etc. In the context of automated systems, most short-term decisions, with the eventual exception of scheduling, are made by the system's control devices.

The study of logistics systems is inherently complex. Despite the numerous models and solutions already developed, the set of constraints and specificities that characterize these systems (availability of resources, random phenomena, interactions between the components, etc.) are not all taken into consideration. It is very often impossible to model such a system in its entirety; hence, in most cases, the provided solutions are not optimal.

The study of logistics systems generally involves several phases. The objective of the first phase is the acquisition of information, which consists of identifying the different parameters influencing the system. The second, very important, phase concerns the development of a model that must integrate all of the identified parameters. The third phase addresses the evaluation and analysis of the system's performance using the initial model. This performance is understood through its quantitative and qualitative aspects. However, the model may be large and difficult to use. An additional simplification phase may then reduce the order of the system, while retaining the characteristics and properties of the initial model. Finally, an optimization phase is regularly used to improve the performance of the system.

The aim of this book is, first, to focus specifically on techniques for modeling logistics systems and to present some single and multiple criteria optimization tools that can be used to solve decision problems, design and management. Second, some new approaches and methods are presented to provide new answers and solutions to some of these problems, such as the design of production lines, their layout, production planning, quality management, stocking and scheduling. Numerous examples are also presented.

The book is organized as follows. Chapter 1, entitled "Modeling and Performance Evaluation", introduces modeling tools such as Markov chains, Petri nets, expansion methods and discrete-event simulation. Chapter 2 discusses optimization methods such as mathematical programming and exact methods (branch-and-bound procedures and dynamic programming). Approach methods and multi-objective optimization are also discussed in the chapter. Chapter 3 deals with the design of logistics systems and their arrangement. It introduces the notions of line-balancing, buffer sizing, and the selection of production equipment. The physical layout of production lines is also introduced. Chapter 4 focuses on tactical management, introducing notions of production planning, replenishment, stock management and quality management. Operational management of production scheduling is the subject of Chapter 5.

Chapter 1

Modeling and Performance Evaluation

1.1. Introduction

A system, be it logistic or otherwise, may be considered as a set of interacting entities, capable of handling other entities that are internal or external to it. A model of a system is a logical, mathematical representation of its real behavior in a given context and following a given problem. A model is a decision-support tool that allows the study of a complex system through one or more simpler systems, replacing it in a scientific analysis, providing information about the studied system or, for example predicting the behavior of the initial system in various conditions.

We may distinguish between different types of model based on different characteristics. A model may be analogical (such as a scale model of a machine) or abstract (without physical representation). If time is not considered in the study, the model is static, and it is described as dynamic if the state of the system it represents evolves over time. If its evolution involves an element of chance, it is described as stochastic (as opposed to deterministic). The notions of deterministic and stochastic models are directly related to uncertainties. These uncertainties are, for example variations in operating times, variations in machine preparation time, etc. If the model requires a formal equation, it is described as mathematical, and it will be called numerical if it is based on a simulation.

The aim of modeling is therefore to best reproduce the actual operation of the studied system. Hence, the physical and technical characteristics of the system must be taken into consideration. This information constitutes a model's input data, and allows the determination, in the most precise possible manner, of output data known as the performance indicators of the system.

In this Chapter, we present a non-exhaustive list of methods and tools for modeling and evaluating the performance of logistics systems, such as Markov chains, Petri nets, the Gershwin decomposition method and discrete-event simulation.

1.2. Markovian processes

Probability theory is a mathematical science that began in the 17th Century with the work of Galileo on physical measurement errors. It was only later in the 19th Century that A. Markov (1856–1922) defined the basis of the theory of stochastic processes, creating the model that carries his name. Markov chains occupy an important place among stochastic models and are currently used in numerous applications (economic, meteorological, military, computational, etc.). Their use in production systems is significant (queuing networks, maintenance policy, fault detection, etc.). A Markov model is well adapted to the study and analysis of production systems because it provides a simple graphical representation while retaining its powerful analytical properties.

1.2.1. *Overview of stochastic processes*

Numerous authors have introduced stochastic processes in their works. We cite the work of [FEL 68].

DEFINITION 1.1.– *A stochastic process is a family of random variables ξ_t:*

$$\{\xi_t, t \in T\} \qquad [1.1]$$

where the parameter t explores the set T. T may belong to the set of natural numbers, real numbers, etc.

If T is discrete and countable: $\{\xi_t\}$ form a stochastic sequence.

If T is a finite or infinite interval: $\{\xi_t\}$ form a continuous process.

Note that T typically represents an interval of time.

Let X be the state space. X is the set whence the variables ξ_t take their values. X may be discrete or continuous.

From the definition of random processes [1.1], we distinguish between four Markov processes according to the discrete or continuous nature of the state space and of the set T (see Table 1.1).

	T discrete	T continuous
X discrete	Markov chains with a discrete state space	Markov processes with a discrete state space
X continuous	Markov chains with a continuous state space	Markov processes with a continuous state space

Table 1.1. *The different Markov processes*

1.2.2. *Markov processes*

The analysis of certain systems in the field, for example, of reliability analysis, imposes continuous-time modeling. We use Markov processes when the transition probabilities at a precise time are replaced by transition rates equivalent to a number of events per unit time. These characteristics are essential as they allow the direct introduction of failure rates and machine repairs into the model.

In this section, we present the basic notions of Markov processes and their applications to production systems modeling.

1.2.2.1. *Basics*

DEFINITION 1.2.– *A Markov process is a family of random variables* $\{\xi_t\}$*, whose parameter belongs to a finite or an infinite continuous interval. The set* T *with index* t *is continuous in this case. The family* $\{\xi_t\}$ *has the Markov property, that is:*

$$Pr\{\xi_t \leqslant x/\xi_{t1} = 1, \xi_{t2} = 2, \ldots, \xi_{tn} = n\} = Pr\{\xi_t \leqslant x/\xi_{tn} = n\} \qquad [1.2]$$

$$\text{with} \quad t > t_n > ... > t_2 > t_1$$

The distribution of the process at a future instant depends only on its present state, and not on its past evolution. The process is memoryless.

For such processes, at each instant, the transition probability from a state i to a state j is zero. Instead of a transition probability π_{ij}, we consider the transition rate μ_{ij}, defined as the limit of the ratio of the transition probability from state i to state j in a small interval Δt from the instant t, with the length of this interval when it tends to zero:

$$\mu_{ij}(t) = \lim_{h \to 0} Pr\{\xi_{t+h} = j/\xi_t = i\} \qquad [1.3]$$

The transition rate may be constant (μ_{ij} = constant) or time-dependent ($\mu_{ij} = \mu_{ij}(t)$). In the former case, the Markov process (with discrete states and continuous time) is said to be homogeneous.

1.2.2.2. *Chapman–Kolmogorov equations*

Let us consider the case of a homogeneous Markov process that possesses a finite number of states $(1, 2, \ldots, r)$. To describe the random process that takes place in this system, state probabilities $(\pi_1(t), \pi_2(t), ..., \pi_r(t))$ are used, where $\pi_i(t)$ is the probability that the system is found in the state i at time t:

$$\pi_i(t) = Pr\{\xi_t = i\} \qquad [1.4]$$

It is clear that for all t, the state probabilities satisfy the property of stochasticity:

$$\sum_{i=1}^{r} \pi_i(t) = 1 \qquad [1.5]$$

To calculate the state probabilities [1.4], it is necessary to resolve the system of differential equations, known as the Chapman–Kolmogorov equations:

$$\frac{d\pi_i(t)}{dt} = \sum_{j=1}^{r} \mu_{ji}.\pi_j(t) - \pi_i(t).\sum_{j=1}^{r} \mu_{ij} \qquad (i = 1, 2, \ldots, r) \qquad [1.6]$$

where $\left(\frac{d\pi_i(t)}{dt}\right)$, also written $\dot{\pi}_i(t)$, corresponds to the rate of variation of the probability associated with state i.

These differential equations [1.6] constitute the state equations of the system. Using matrices, we obtain:

$$[\dot{\pi}_1(t), \dot{\pi}_2(t), ..., \dot{\pi}_r(t)] = [\pi_1(t), \pi_2(t), ..., \pi_r(t)] \ . \ A \qquad [1.7]$$

$$\text{with} \quad \pi(t) \in \Re^{1\times r}, \ A \in \Re^{r\times r} \qquad [1.8]$$

Matrix A is called the transition rate matrix, or in certain works, the infinitesimal generator of the Markov process.

a_{ij} is the transition rate μ_{ij} of state i to state j $(i \neq j)$. a_{ii} is equal to the negative of the sum of transition rates from state i to all other states.

The elements of A have the following properties:

$$\begin{cases} a_{ij} \geqslant 0 & i \neq j \\ a_{ii} = -\sum\limits_{\substack{j=1 \\ i\neq j}}^{n} a_{ij} & \end{cases} \qquad [1.9]$$

The sum of the elements in each line is zero and the determinant of A is therefore zero. Equation [1.6] may be written in a more condensed form with $\dot{\pi}(t)$, the derivative of the state probability vector at time t:

$$\dot{\pi}(t) = \pi(t) \ . \ A \qquad [1.10]$$

This equation perfectly defines the transient state of the system. To compute the probability distribution $\pi(t)$ at every instant t, it is necessary to solve the system of differential equations [1.6]. When the Markov process is homogeneous, the solution of the system is expressed in the form:

$$\pi(t) = \pi(0) \ . \ e^{A.t} \qquad [1.11]$$

where $\pi(0)$ is the state probability vector at time $t = 0$.

1.2.2.3. *Steady-state probabilities*

The existence of limiting state probabilities requires that the Markov process be ergodic. In other words, the limit of $Pr\{\xi_t \leqslant x\}$ when t tends to infinity exists regardless of the initial distribution. Thus:

$$\lim_{t \to \infty} \pi_i(t) = \pi_i(\infty) = \pi_i^* \qquad i = 1, 2, \ldots, r \qquad [1.12]$$

In the limiting case, the variation of the state probabilities $\dot{\pi}(t)$ is zero. Equation [1.6] becomes:

$$\sum_{j=1}^{r} \mu_{ji}.\pi_j(t) = \pi_i(t). \sum_{j=1}^{r} \mu_{ij} \qquad (i = 1, 2, \ldots, r) \qquad [1.13]$$

The quantity $\mu_{ij}.\pi_i(t)$ is called the probability flow of the transition of state i to state j.

Equation [1.13] is called the balance equation. This corresponds to an equilibrium between the total probability flow entering state i and the one leaving it.

To solve the system of linear equations, it is necessary to add an equation corresponding to the stochasticity of the steady-state probabilities:

$$\begin{cases} \pi(\infty) \ . \ A = 0 \\ \displaystyle\sum_{i=1}^{r} \pi_i(\infty) = 1 \end{cases} \qquad [1.14]$$

The solution to equation [1.14] gives us the limiting distribution of the state probabilities:

$$\pi(\infty) = [\pi_1(\infty), \pi_2(\infty), \ldots, \pi_r(\infty)] \qquad [1.15]$$

1.2.2.4. *Graph associated with a Markov process*

An oriented graph $G = (X, E)$ is associated with a Markov process, where X is the set of vertices of the graph, which are the states, and E is the set of oriented arcs linking each vertex. The graph is weighted by the transition rates μ_{ij} associated with the arcs. As a general rule, and in contrast to Markov chains, the transition rates μ_{ii} are not represented on the graph.

1.2.2.5. *Application to production systems*

The observation of a workstation in a workshop shows that over time, each element of the productive activity moves successively in a deterministic or random way through a number of different phases referred to as elementary states. These elements are categorized into three groups that are found in every production facility:

– the products;

– the means of production;

– the production operators.

Generally, we may say that each element (product, means, and operator) is successively found in one of the two primary states according to whether or not they are participating in the production process. We may call these elementary states:

– an operating or active state (participation in production);

– an idle or inactive state (non-participation in production).

Random phenomena that disrupt the proper operation of the production facility depend on different elements, for example:

– products: absence, defect, etc.;

– means: failure (mechanical, electrical, hydraulic, etc.), waiting, calibration, repair, etc.;

– operator: pause, absence, incident, etc.

The different elementary states of the production system are modeled using Markov processes, where each state corresponds to a vertex on the graph. Next, the different links that exist between the vertices of the graph must be determined. Each link is assigned a rate of transition between states per unit time. This rate may correspond, for example to a machine's rate of failure or repair. This characteristic makes Markov processes crucial in operations research, as they serve as a starting point for the development of several models of queuing, renewal, equipment maintenance and stock management. Here is an example of a production system model.

EXAMPLE 1.1.– Let us consider a production cell composed of two workstations: a manufacturing machine and an assembly machine.

The behavior of the cell may be decomposed into four elementary states:

– state 1: production;

– state 2: breakdown of the manufacturing machine;

– state 3: breakdown of the assembly machine;

– state 4: breakdown of the cell (the manufacturing and assembly machines).

The transition rates between the different states are determined from the failure and repair rates of the two machines. The numerical values of these two rates are given in Table 1.2. Figure 1.1 shows the graph of the Markov process associated with the production cell.

Machine	Failure rate (per month)	Repair rate (per month)
Manufacturing	$\lambda_1 = 1$	$\mu_1 = 10$
Assembly	$\lambda_2 = 5$	$\mu_2 = 11$

Table 1.2. *Characteristics of the machines*

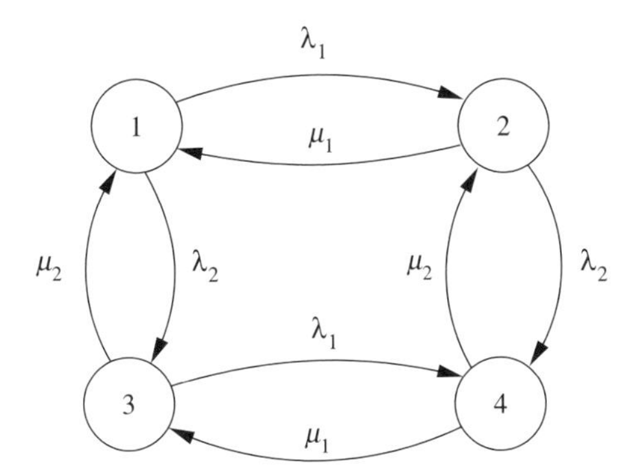

Figure 1.1. *Graph of the Markov process associated with the production cell*

From this graph, the generator A that contains the set of transition rates between the four states is formulated:

$$A = \begin{bmatrix} -(\lambda_1 + \lambda_2) & \lambda_1 & \lambda_2 & 0 \\ \mu_1 & -(\mu_1 + \lambda_2) & 0 & \lambda_2 \\ \mu_2 & 0 & -(\lambda_1 + \mu_2) & \lambda_1 \\ 0 & \mu_2 & \mu_1 & -(\mu_1 + \mu_2) \end{bmatrix} \qquad [1.16]$$

$$A = \begin{bmatrix} -6 & 1 & 5 & 0 \\ 10 & -15 & 0 & 5 \\ 11 & 0 & -12 & 1 \\ 0 & 11 & 10 & -21 \end{bmatrix}$$

The transient state of the system is given by the eigenvalues of the generator A:

$$\lambda(A) = \begin{bmatrix} 0 & -11 & -16 & -27 \end{bmatrix}$$

From the system of linear equations [1.14], we calculate the steady-state probabilities:

$$\pi(\infty) = \begin{bmatrix} \pi_1(\infty) & \pi_2(\infty) & \pi_3(\infty) & \pi_4(\infty) \end{bmatrix}$$

$$\pi(\infty) = \begin{bmatrix} 0.625 & 0.0625 & 0.2841 & 0.0284 \end{bmatrix}$$

Let us consider that at the initial time $t = 0$, the system is found in the production state (state 1). Thus, $\pi(0) = [1\ 0\ 0\ 0]$. The evolution of the state probabilities is given in Figure 1.2.

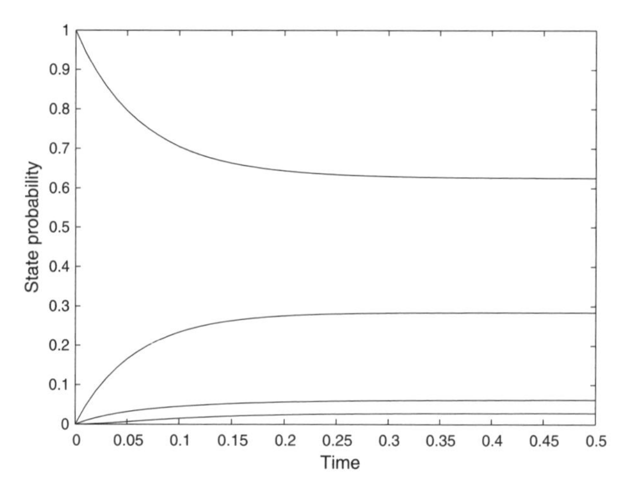

Figure 1.2. *Evolution of the state probabilities*

1.2.3. *Markov chains*

1.2.3.1. *Basics*

DEFINITION 1.3.– *A Markov chain $\{\xi_n\}$ is a random Markov process whose state space X is a finite or an infinite set. ξ_n may be considered as the result of the test n.*

Generally, the state space X is identified by non-negative integers $(0, 1, 2, \ldots)$. If the result of test n is state i, we write it $\xi_n = i$. The probability that the process is in state j at time n given that it was in state i at time $(n-1)$ is written as $\pi_{ij}^{n-1,n}$. This is called the transition probability:

$$\pi_{ij}^{n-1,n} = Pr\{\xi_n = j/\xi_{n-1} = i\} \qquad [1.17]$$

The Markov property states that the future behavior of the sequence of events depends exclusively on the knowledge of the present state. Thus:

$$Pr\{\xi_n = j/\xi_{n-1} = i, \dots, \xi_0 = a\} = Pr\{\xi_n = j/\xi_{n-1} = i\} \quad [1.18]$$

DEFINITION 1.4.– *The Markov chain is said to be homogeneous in time if its transition probabilities are independent of the considered time n. The transition probabilities are then stationary, and satisfy:*

$$Pr\{\xi_n = j/\xi_{n-1} = i\} = \pi_{ij} \quad [1.19]$$

These transition probabilities satisfy the following properties:

$$\sum_{j=1}^{r} \pi_{ij} = 1 \qquad i = 1, 2, \dots, r \quad [1.20]$$

and

$$0 \leq \pi_{ij} \leq 1 \qquad i, j = 1, 2, \dots, r \quad [1.21]$$

1.2.3.2. *State probability vectors*

Let $\pi_i(n)$ be the probability that the system is in state i at time n:

$$\pi_i(n) = Pr\{\xi_n = i\}, \qquad n = 0, 1, 2, \dots \quad [1.22]$$

Grouping these probabilities into a vector, called the state probability vector at time n, we obtain:

$$\pi(n) = [\pi_1(n), \pi_2(n), \dots, \pi_r(n)] \quad [1.23]$$

At the initial time ($t = 0$), this probability vector is written as:

$$\pi(0) = [\pi_1(0), \pi_2(0), \dots, \pi_r(0)] \quad [1.24]$$

The state probabilities present the following properties:

$$\sum_{i=1}^{r} \pi_i(n) = 1 \quad [1.25]$$

and

$$0 \leq \pi_i(n) \leq 1 \qquad i = 1, 2, \dots, r \qquad n = 0, 1, 2, \dots \quad [1.26]$$

1.2.3.3. *Fundamental equation of a Markov chain*

The state probabilities at time $(n + 1)$ are given by the law of total probability:

$$\pi_i(n+1) = \pi_1(n).\pi_{1i}(n) + \pi_2(n).\pi_{2i}(n) + ... + \pi_r(n).\pi_{ri}(n) \quad [1.27]$$

$$= \sum_{j=1}^{r} \pi_j(n).\pi_{ji}(n) \qquad i = 1, 2, ..., r \quad [1.28]$$

or in vectorial form:

$$\pi_i(n+1) = [\pi_1(n), \pi_2(n), \ldots, \pi_r(n)]. \begin{bmatrix} \pi_{1i}(n) \\ \pi_{2i}(n) \\ \vdots \\ \pi_{ri}(n) \end{bmatrix} \qquad i = 1, 2, \ldots, r \quad [1.29]$$

We arrive at the matrix equation, called the fundamental equation, of a Markov chain:

$$\pi(n+1) = \pi(n).\tau(n) \qquad [1.30]$$

$$\text{with} \quad \tau(n) = \begin{bmatrix} \pi_{11}(n) & \pi_{12}(n) & \cdots & \pi_{1r} \\ \pi_{21}(n) & \pi_{22}(n) & \cdots & \pi_{2r} \\ \vdots & \vdots & \vdots & \vdots \\ \pi_{r1}(n) & \pi_{r2}(n) & \cdots & \pi_{rr} \end{bmatrix} \qquad [1.31]$$

This equation gathers all the information necessary to pass from the state probability distribution at time n to the state probability distribution at time $n+1$.

The matrix τ is called the transition matrix at time n. It gathers together the set of transition probabilities. This matrix is stochastic because its elements satisfy properties [1.20] and [1.21].

In the case of a homogeneous Markov chain, equation [1.30] is written as:

$$\pi(n+1) = \pi(n).\tau \qquad [1.32]$$

As a result, if $\pi(0)$ is the initial distribution of the state probabilities, the distribution at later times will be as in the case of a homogeneous Markov chain:

$$\pi(1) = \pi(0).\tau$$

$$\pi(2) = \pi(1).\tau = \pi(0).\tau^2$$

$$\pi(3) = \pi(2).\tau = \pi(0).\tau^3$$

$$\vdots$$

$$\pi(n) = \pi(n-1).\tau = \pi(0).\tau^n \qquad [1.33]$$

Formula [1.33] allows us to define the asymptotic properties of τ, that is it allows us to calculate the steady-state probabilities.

1.2.3.4. *Graph associated with a Markov chain*

With each Markov chain, we associate an oriented graph $G = (X, E)$, where X represents the set of states, i.e. the vertices of the graph, and E represents the set of arcs of the graph connecting each vertex.

The graph may be weighted by the transition probabilities π_{ij} associated with the arcs connecting state i to state j.

The graphical model of Markov chains is based, of course, on graph theory to obtain its properties, such as the asymptotic properties of $\pi(n)$.

1.2.3.5. *Steady states of ergodic Markov chains*

First, let us define the ergodicity property.

PROPERTY 1.1.– A Markov chain is ergodic if, in its asymptotic behavior, the system tends to a unique limit in distribution, independent of the initial conditions:

$$\pi(\infty) = \lim_{n \to \infty} \pi(n) = \pi^* \qquad \forall \quad \pi(0) \tag{1.34}$$

The study of state probabilities in ergodic, steady-state Markov chains is mainly carried out using one of two methods: one direct and the other indirect. The indirect method consists of writing the transition matrix τ in diagonal form using a modal transformation. Here, we develop the direct, simplest and, above all, fastest method.

When we make n tend to infinity in the fundamental equation [1.32], we obtain:

$$\pi(\infty) = \pi(\infty) \, . \, \tau \tag{1.35}$$

then:

$$\pi(\infty) \, . \, (\tau - I) = 0 \tag{1.36}$$

We write $D = (\tau - I)$, where I is an $r \times r$-dimensional unit matrix.

Equation [1.36] is then written:

$$\pi(\infty) \, . \, D = 0 \tag{1.37}$$

The matrix D is known as the dynamic matrix.

Using the stochasticity property of the state probability vector, the linear system of equations is written:

$$\begin{cases} \pi(\infty) \, . \, D = 0 \\ \pi_1(\infty) + \pi_2(\infty) + \ldots + \pi_r(\infty) = 1 \end{cases} \tag{1.38}$$

Solving the above equation gives us the limit in distribution of the state probabilities.

$$\pi(\infty) = [\pi_1(\infty), \pi_2(\infty), \ldots, \pi_r(\infty)] \tag{1.39}$$

1.2.3.6. *Application to production systems*

The representation of production systems in the form of state models corresponds exactly to the model offered by Markov chains because each vertex of the graph represents a state of the system. It suffices to determine the transition probabilities between the different states.

Generally, numerous industrial examples have been studied using Markov chains, such as queuing networks and problems in operational research.

It is undeniable that Markov chains are very efficient tools for the study of production systems.

In addition to a simple graphical model, the mathematical model associated with Markov chains allows the calculation of numerous performance indices, such as reliability, availability and even the maintainability of a system.

EXAMPLE 1.2.– During its life, each piece of equipment passes through different degrees of wear. We associate a state in a Markov chain with each of these degrees of wear.

With each instance of repair or maintenance work carried out, the equipment moves to a lower degree of wear. Let us take as an example a production machine (lathe, milling machine, etc.) having four degrees of wear. The transition matrix takes the following form:

$$\mathcal{T} = \begin{bmatrix} 0.6 & 0.2 & 0.1 & 0.1 \\ 0.2 & 0.4 & 0.3 & 0.1 \\ 0 & 0.2 & 0.4 & 0.4 \\ 0.1 & 0 & 0.4 & 0.5 \end{bmatrix} \qquad [1.40]$$

From this transition matrix, we formulate the state graph of the Markov chain (Figure 1.3).

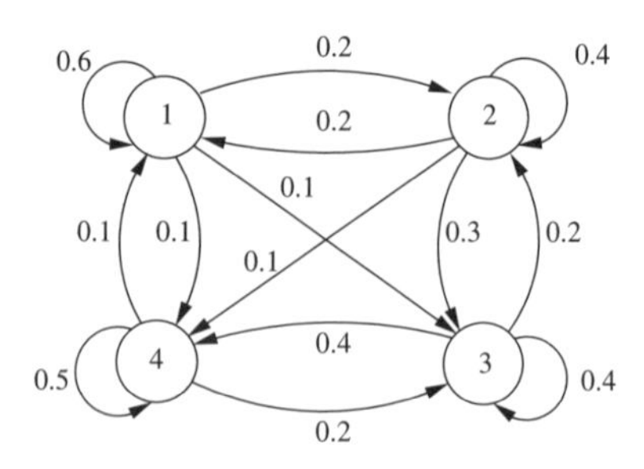

Figure 1.3. *The Markov chain associated with a system of wear*

With each degree of wear, i, we associate a state i of the graph. The link between states correspond to the transition probabilities between the degrees of wear.

We consider the equipment to be initially in state 1. Thus, $\pi(0) = [1\ 0\ 0\ 0]$.

Using the fundamental equation [1.32], we may calculate the state probabilities at different times n.

Figure 1.4 represents this evolution:

$$\pi(1) = \pi(0)\ .\ \tau = [1\ 0\ 0\ 0]\ .\ \begin{bmatrix} 0.6 & 0.2 & 0.1 & 0.1 \\ 0.2 & 0.4 & 0.3 & 0.1 \\ 0 & 0.2 & 0.4 & 0.4 \\ 0.1 & 0 & 0.4 & 0.5 \end{bmatrix} = [0,6 \quad 0,2 \quad 0,1 \quad 0,1]$$

To calculate the steady-state probabilities, we solve the system of linear equations [1.41], which consists of four equations with four variables:

$$\begin{cases} \pi(\infty)\ .\ D = 0 \\ \pi_1(\infty) + \pi_2(\infty) + \pi_3(\infty) + \pi_4(\infty) = 1 \end{cases} \qquad [1.41]$$

Thus, we obtain $\pi(\infty) = \left[\frac{1}{6} \quad \frac{1}{6} \quad \frac{1}{3} \quad \frac{1}{3}\right]$.

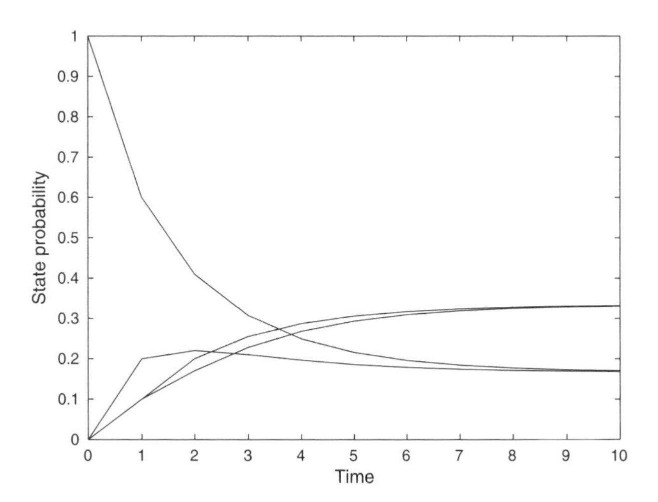

Figure 1.4. *The evolution of the state probabilities*

In steady state, in one sixth of the time, the system is in state 1 or in state 2 and in one third of the time the system is in states 3 or 4.

1.3. Petri nets

A Petri net (PN) is a graphical and mathematical modeling tool, which is used in a large number of domains such as automation, operations management and computer science. It allows the description of events and simultaneous, shared and synchronized evolution.

Petri nets were created by Carl Adam Petri in 1962. Numerous research teams currently work with this tool, developing simulation software for different applications such as production systems and information systems.

1.3.1. *Introduction to Petri nets*

In this section, we recall the principal definitions of autonomous Petri nets and their particular structures, explaining the static and dynamic aspects as well as their structural properties.

1.3.1.1. *Basic definitions*

DEFINITION 1.5.– *A Petri net is a bipartite-oriented graph whose nodes are places and transitions.*

A Petri net may be defined as a four-tuple:

$$PN = (P, T, Pre, Post) \quad [1.42]$$

where

– P *is the finite set of places,* $P = (P_1, P_2, \ldots, P_n)$*.*

– T *is the finite set of transitions,* $T = (T_1, T_2, \ldots, T_m)$*.*

– *Pre:* $(P \times T) \to \mathbb{N}$ *is the forward incidence mapping.*

– *Post:* $(P \times T) \to \mathbb{N}$ *is the backward incidence mapping.*

– *The matrices* Pre*,* $Post$ *are* $n \times m$ *matrices.*

– $Pre(P_i, T_j)$ *is the weight of the arc linking* P_i *to* T_j*.*

– $Post(P_i, T_j)$ *is the weighting of the arc linking* T_j *to* P_i*.*

Graphically, places are represented by circles, and transitions are represented by bars.

The arcs linking places to transitions and transitions to places are oriented.

There is always an alternation between places and transitions (two places or two transitions cannot be linked) as in the example in Figure 1.5.

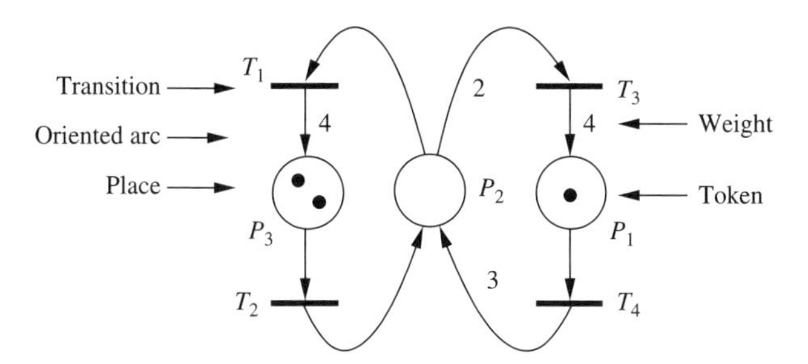

Figure 1.5. *An example of a Petri net*

Notation:

– $^\circ T_j$ (T_j°) is the set of input (output) places of the transition T_j.

– $^\circ P_i$ (P_i°) is the set of input (output) transitions of the place P_i.

The incidence matrix W is defined by:

$$W = Post - Pre \qquad [1.43]$$

REMARK 1.1.– $P \cap T = \varnothing$ and if Pre and $Post \in \{0, 1\}$, then the Petri net is ordinary.

DEFINITION 1.6.– *A marked Petri net is a pair* (PN, M)*, where:*

– $PN = (P, T, Pre, Post)$ *is a Petri net.*

– $M : P \to \mathbb{N}$ *is a marking function.*

The initial marking of the Petri net is written as M_0. The marking of the Petri net is carried out by tokens. Graphically, the tokens (or marks) are represented by black disks and are placed in the center of the places.

1.3.1.2. *Dynamics of Petri nets*

The dynamics of Petri nets are obtained by inserting tokens or marks into places. The tokens will thus circulate from place to place by crossing different alternating transitions. The crossing (or "firing") of a transition must follow certain rules. A transition T_j may fire (be crossed) for a marking M if and only if:

$$M(P_i) \geq Pre(P_i, T_j), \ \forall P_i \in {}^\circ T_j$$

The firing of a transition T_j consists of taking $Pre(P_i, T_j)$ tokens from each place $P_i \in {}^\circ T_j$ and adding $Post(P_i, T_j)$ tokens from every place $P_i \in T_j^\circ$. Thus, from marking M, the firing of a transition T_j leads to a new marking M' defined as follows:

$$M'(P_i) = M(P_i) - Pre(P_i, T_j) + Post(P_i, T_j), \ \forall P_i \in P$$

A crossable transition does not necessarily immediately fire, and only one firing is possible at a given moment. Figure 1.6 shows the firing of transition T_1. In this example, the number of tokens in the Petri net is not conserved. Indeed, the total number of tokens is four before and five after T_1 fires. The marking obtained after T_1 fires is the following:

$$M' = M - Pre + Post$$

hence:

$$\begin{bmatrix}2\\2\\0\\0\\0\end{bmatrix} - \begin{bmatrix}2\\1\\0\\0\\0\end{bmatrix} + \begin{bmatrix}0\\0\\2\\2\\1\end{bmatrix} = \begin{bmatrix}0\\1\\2\\2\\1\end{bmatrix}$$

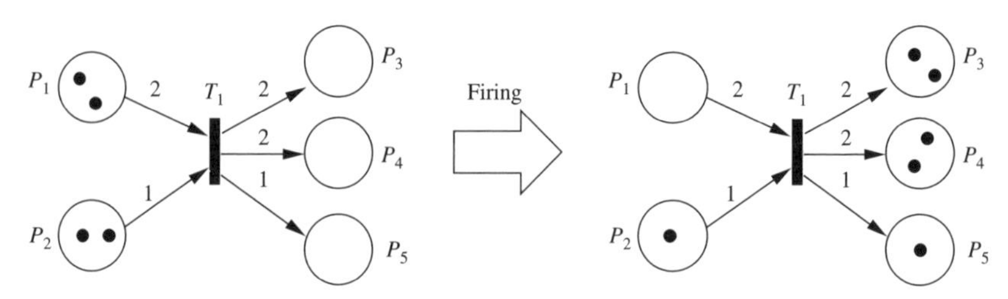

Figure 1.6. *The firing of a transition*

1.3.1.3. *Specific structures*

A loop (P_i, T_j) is such that $P_i \in {}^\circ T_j$ and $P_i \in T_j^\circ$ (Figure 1.7(a)). In the incidence matrix W, there is a loss of information regarding the loop.

A source transition is a transition without an input place (it may always fire). An example of a source transition source is given in Figure 1.7(b). A sink transition is, for its part, a transition without an output place. An example of a sink transition is given in Figure 1.7(c). Loops are very important in production systems modeling because they limit the number of products that a machine may process simultaneously.

Two transitions T_1 and T_2 are in structural conflict if and only if they have at least one input place P_i in common. An example of structural conflict is shown in Figure 1.8(a). Two transitions T_1 and T_2 are in effective conflict for a marking M if and only if they have at least one input place P_i in common such that:

$$M(P_i) \geq Pre(P_i, T_1) \qquad \text{and} \qquad M(P_i) \geq Pre(P_i, T_2)$$

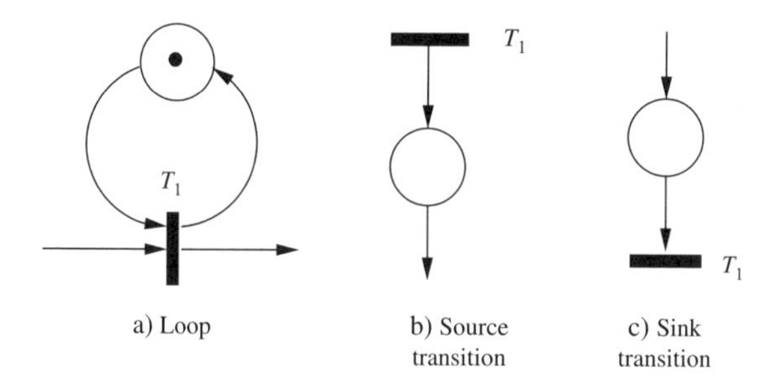

Figure 1.7. *Some specific structures: a loop, a source, and a sink*

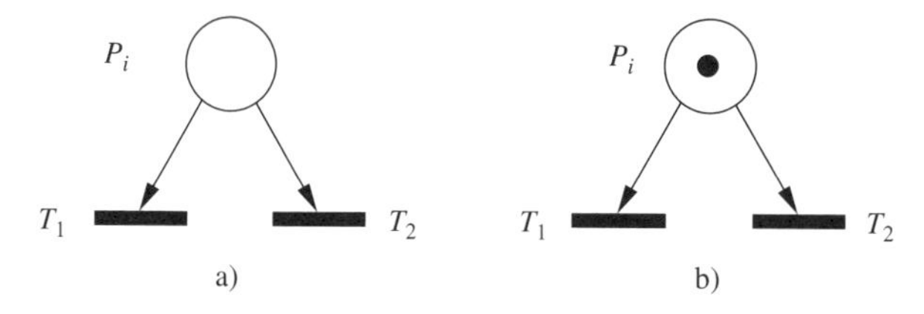

Figure 1.8. *Examples of (a) structural and (b) effective conflict*

An example of effective conflict is shown in Figure 1.8(b).

Conflicts are also found in production systems modeling. In a scheduling problem, they allow the modeling of the choice between two tasks to be carried out, or the choice between two machines.

Inhibitor arcs make possible the crossing of transitions when places are empty or when they have a marking of less than a set value. Figure 1.9 shows that the transition T_1 may fire if there is a minimum of one token in P_1 and a maximum of three tokens in P_3. The inhibitor arc linking P_3 to T_1 works in the opposite way to a normal arc. Note that an inhibitor arc always connects a place to a transition, and not vice versa.

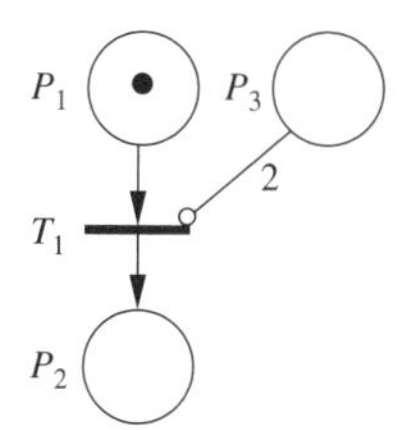

Figure 1.9. *An example of a Petri net with an inhibitor arc*

In Figure 1.9, from the marking $M_0 = \{1, 0, 0\}$, we obtain the marking $M_0 = \{0, 1, 0\}$ after T_1 fires.

REMARK 1.2.– An inhibitor arc does not remove tokens from its input place.

Inhibitor arcs are also very important in the study of production systems. They are widely used to model inventory policies because they can trigger replenishment when the stock level is less than a certain value.

1.3.1.4. *Tools for Petri net analysis*

The main methods of Petri net analysis may be classified into three groups:

– methods based on the enumeration of all the accessible states: marking trees and graphs and covering trees and graphs;

– methods based on linear algebra;

– Petri net reduction methods.

The tree of reachable markings defines all the markings that may be attained starting from M_0. This is a tree in which:

– The nodes are the markings attainable from M_0.

– Each arc represents the firing of a transition.

– The root of the tree corresponds to M_0.

The reachability graph is deduced from the reachability tree by merging the nodes with the same marking. Note that the number of reachable markings may be infinite (the non-boundedness property); in this case, it is impossible to build the reachability tree.

The spanning tree (or covering tree) limits the size of the tree when the number of states is not finite. For this, a new symbol ω is introduced, which corresponds to infinity. The state equation is defined by:

$$M' = M + W.\underline{s} \qquad [1.44]$$

where M is the initial marking, $\underline{s}$ is the characteristic vector of the firing sequence s, and whose j-component corresponds to the number of firings of transition T_j in the sequence s, and M' is the new marking obtained from M after the firing sequence s has been executed.

Note that the state equation does not guarantee the firability of the sequence s (i.e. whether or not the sequence may be realized), but only gives the final marking obtained after a firing sequence.

This state equation may be used without much difficulty in the creation of a computer program (in Excel, Matlab®, C or other programming languages) to simulate the behavior of an autonomous Petri net.

The following algorithm allows the step-by-step, firing-by-firing simulation of the evolution of a Petri net. This algorithm guarantees that at each step, the transition firing sequence is valid.

At the end of the simulation, we may simply trace the evolution of the markings on a graph. Here, we are concerned with discrete-event simulation with Petri nets.

Algorithm 1.1 Autonomous Petri net simulation algorithm

Data : $M_0, Pre, Post$

Repeat

STEP 1: From the current marking M, seek the enabled transitions. Hence deduce the characteristic firing vector $\underline{s}$.

STEP 2: Fire the enabled transitions and determine the new marking M' using the relation: $M' = M + W.\underline{s}$.

Until The number of firings is not reached.

1.3.1.5. *Properties of Petri nets*

1.3.1.5.1. Reachability

DEFINITION 1.7.– *The determination of the reachability of a marking M' from a marking M consists of verifying that there is a sequence of transitions s that allows the marking M to be attained from the marking M'.*

1.3.1.5.2. Boundedness

DEFINITION 1.8.– *A place P_i in a marked Petri net $\langle PN, M_0 \rangle$ is k-bounded if there exists a positive integer k such that for every marking M that is reachable from the initial marking M_0, we have:*

$$M(P_i) \leq k$$

A place P_i is bounded if there exists a positive integer k such that P_i is k-bounded.

A marked Petri net $\langle PN, M_0 \rangle$ is k-bounded if all of its places are k-bounded.

A Petri net is bounded if there exists a positive integer k such that it is k-bounded. The net is said to be safe if it is 1-bounded.

1.3.1.5.3. Liveness and blocking

DEFINITION 1.9.– *A transition T_j of a Petri net $\langle PN, M_0\rangle$ is live if, for any marking M that is accessible from the marking M_0, there exists a firing sequence, which from M, leads to a marking that allows the transition T_j to fire.*

DEFINITION 1.10.– *A Petri net $\langle PN, M_0\rangle$ is live if all its transitions are live.*

1.3.1.5.4. Application to production systems

These various properties have applications to production systems. In fact, the boundedness property implies a lack of increase in the number of parts in the system. This may correspond to a finite capacity for a stock. The boundedness of the net ensures that the stock reaches a certain level without ever going over it. The liveness property indicates that there are no sinks. It also guarantees that the production system can produce correctly. There is no risk of stopping after a certain operating time.

1.3.2. *Non-autonomous Petri nets*

In a non-autonomous Petri net, evolution depends not only on the state of the net, but also on the environment in which it is included. Thus, the evolution may be time-dependent. Several Petri nets such as timed, temporal, stochastic and continuous Petri nets, take this into consideration. This evolution may also depend on external data, and Petri nets that integrate these data are called high-level Petri nets. Examples of this category include colored, object or predicate-transition Petri nets. The following sections give the definitions and operation of the main Petri nets. More detail is given regarding continuous and stochastic Petri nets, which are efficient models and are widely used in the study of industrial systems.

1.3.3. *Timed Petri nets*

Timed Petri nets allow the description of systems whose operation depends on time (the duration between the start and the end of an operation). The introduction of time may be done either in transitions or in places. Thus, we distinguish between two types of nets: P-timed and T-timed Petri nets. Only T-timed Petri nets are discussed in this book.

DEFINITION 1.11.– *A T-timed Petri net is a pair $\langle PN, Temp\rangle$ with*

PN a marked Petri net:

$$PN = (P, T, Pre, Post, M_0)$$

$Temp$: an application from the set T into $\mathbb{Q}^+$;

$Temp(T_i) = d_i$ = the time delay associated with the transition T_i.

Marks have two states:

– reserved while the firing of the transition is decided;

– not reserved.

The validation of a transition does not imply the immediate reservation of the mark.

At the initial time, all marks are not reserved. The calculation of performance indices is carried out in steady state, i.e. when the evolution of the markings is constant and stable. Three major performance indices emerge readily. By simply counting the number of tokens that reside in the places or that cross the transitions in a fixed interval, we obtain:

– the average number or marks in a place, denoted by M^*;

– the mean firing frequency of the transitions, denoted by f^*;

– the mean waiting time of a mark in a place, denoted by t^*.

If we apply the previously defined performance indices to production systems, we obtain the following:

– the average number of items produced per unit time;

– the mean utilization rate of the machines;

– the average number of items in stock;

– the mean sojourn time of an item in a stock.

EXAMPLE 1.3.– Consider a simple production system composed of a production unit (machine) and a stock. The aim is to study the flow of pallets through the system. We will consider the flow of pallets to be closed. Hence, we consider the transitions as only validating one token at a time (modeling is therefore implicit, and loops are not added to limit the reservation to a single token per firing). The places are of unlimited capacity. The Petri net model is shown in Figure 1.10. The initial marking of the net is therefore $M_0 = \{10, 0\}$. Analyzing the properties of the net, we find that it is bounded (the number of tokens in places P_1 and P_2 is at most 10), live (both transitions T_1 and T_2 are firable from any marking reachable from the initial marking M_0), and reversible (for any marking M reachable from M_0, M_0 is reachable from M). To determine the different markings over time, we use Algorithm 1.2. It is assumed that transitions undergo only one firing at once (no q-validation).

In our case, we are interested in the mean number and mean throughput of pallets in the different parts of the system. Measures of performance (mean place markings and mean firing frequencies of the transitions) are only calculated in the steady state of the system, i.e. on marking evolution curves in the time interval $[10, +\infty[$.

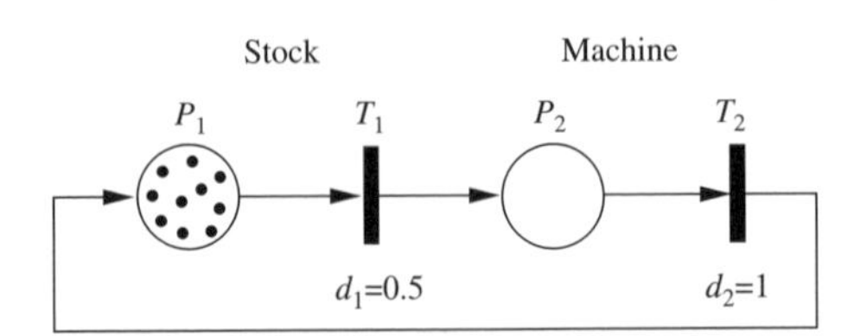

Figure 1.10. *A T-timed Petri net model of the flow of pallets*

Algorithm 1.2 Timed Petri net simulation algorithm

Data: M_0, Pre,$Post$, $delay$ (transition delay vector)
Initialization of the instantaneous firing vector $fire$ and the stopwatch $watch$: $\forall i\ fire(i) = \infty$, $watch = 0$
Repeat
STEP 1: From the current marking M, search the enabled transitions. Deduce the characteristic firing vector $\underline{s}$. For each enabled transition i, carry out the operation: $fire(i) = delay(i) + watch$
STEP 2: Increment the watch to the next firing time: $watch = min(fire(i))$. Fire the enabled transitions and determine the new marking vector M' using the formula: $M' = M + W.\underline{s}$. For each fired transition i, carry out the operation $fire(i) = \infty$.
STEP 3: Record the new marking vector M' as well as the firing time.
Until The number of firings is not reached.
Trace the evolution of markings as a function of time.

Thus, we obtain:

– the average number of tokens in the stock: $M_1^* = 0.5$;

– the average number of tokens in the machine: $M_2^* = 9.5$;

– the mean flow leaving the stock: $f_1^* = 1$;

– the mean flow leaving the machine: $f_2^* = 1$.

The evolution curves clearly show that the marking of P_1 evolves between 0 and 1 and that of P_2 evolves between 9 and 10, with the same sojourn time (0.5 time units) that gives, on average, 0.5 pallets in stock and 9.5 pallets in the machine.

The mean frequencies are identical and equal to one pallet per unit time. In a production line, it is always the slowest component (here the production unit) that gives the mean throughput rate of the whole line.

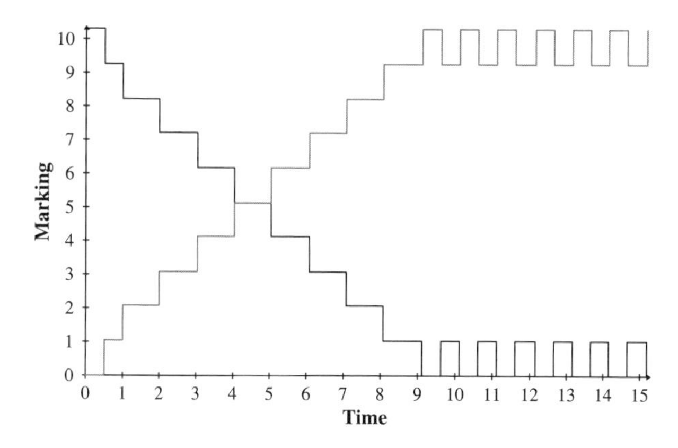

Figure 1.11. *The evolution of the markings*

1.3.4. *Continuous Petri nets*

Continuous Petri nets are used when the number of marks is very significant. It is then impossible to graphically represent all of the marks in the places.

The marks are replaced by an integer and the time delays are replaced by firing speeds. An example of a continuous Petri net is shown in Figure 1.12.

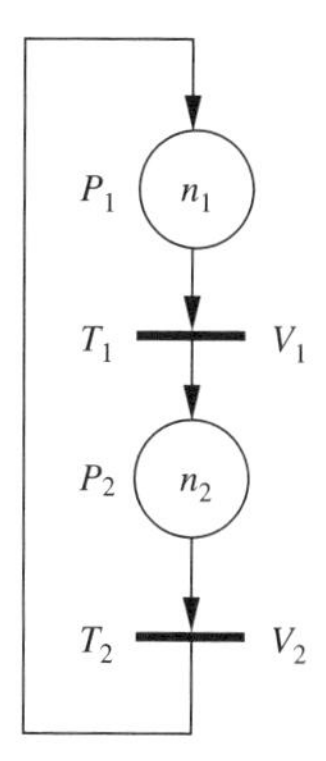

Figure 1.12. *A continuous Petri net*

The operation of a continuous Petri net is initially defined as an extension of the operation of T-timed Petri nets. Tokens are divided into parts until an infinity of sub-tokens is obtained. The time delay associated with a transition is then replaced by a firing speed. This speed corresponds to the quantity of tokens fired per unit time. We may compare the operation of a transition and an upstream place with an hourglass, where the sand grains are tokens. We may also use the analogy of a fluid that varies between zero and infinity in a continuous manner, i.e. a place may have a marking of 0.6. From the notion that we may have a marking of less than 1, two approximations are possible:

– $V_1 = \frac{1}{d_1}$. The Petri net is called a constant speed continuous Petri net (CCPN).

– $V_1 = \frac{1}{d_1}.min(1, m_1)$ or $V_i = \frac{1}{d_i}.min(1, ma, \ldots, mb)$ with $\{P_a, \ldots, P_b\} = {}^{\circ}T_i$. The Petri net is called a variable speed continuous Petri net (VCPN).

In the first approximation, the speed is constant and does not take into account the marking of upstream places. In the second approximation, the speed is variable as it will depend on the marking of upstream places, but only when the marking is less than 1. On the other hand, both types of continuous Petri nets, CCPNs and VCPNs, have exactly the same behavior when $m_1 \geq 1$.

1.3.4.1. *Fundamental equation and performance analysis*

The evolution of a marking in time is given by this fundamental equation:

$$\frac{dM(t)}{dt} = W.v(t)$$

where

– $M(t)$: the marking vector at time t;

– $M(t) = (m_1(t), \ldots, m_n(t))$, where $m_i(t)$ is the marking of place P_i;

– W: the incidence matrix;

– $v(t)$: the firing speed vector at time t;

– $v(t) = (v_1(t), \ldots, v_m(t))$, where $v_i(t)$ is the firing vector of transition T_i.

This fundamental equation gives a system of first-order differential equations. The solution of these equations gives increasing or decreasing linear or exponential evolution.

Performance evaluation with continuous Petri nets is carried out by solving the system of first-order differential equations obtained from the fundamental equation defined in section 1.3.4.1. We then obtain the different equations for the evolution of markings in the net. Note:

– The evolution equations are only valid in well-specified intervals of time.

– Markings are always positive or zero: $\forall i, \forall t, m_i(t) \geq 0$.

– For VCPNs, the firing speed depends on the markings. Performance evaluation is carried out step-by-step or phase-by-phase. A phase change takes place as soon as the evolution of a marking crosses the horizontal line $y = 1$ either by increasing or by decreasing. The marking goes either above or below 1.

1.3.4.2. *Example*

Let us take again the example used for timed Petri nets – the study of a closed flow of pallets in a stock and a production machine. The continuous Petri net model of this system is shown in Figure 1.13. In the case of a CCPN the firing speeds are as follows:

$$\begin{cases} v_1(t) = U_1 = 2 & \text{(constant)} \\ v_2(t) = U_2 = 1 & \text{(constant)} \end{cases}$$

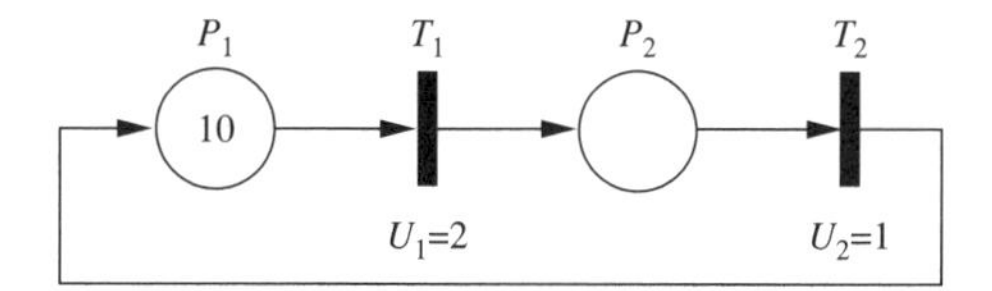

Figure 1.13. *Continuous Petri net model of the flow of pallets*

From the definition of firing speeds, we obtain the marking evolution equations for places P_1 and P_2:

$$\begin{cases} m_1(t) = m_1(0) + (U_2 - U_1).t = 10 - t \\ m_2(t) = m_2(0) + (U_1 - U_2).t = t \\ \forall i, \forall t, m_i(t) \geq 0. \end{cases}$$

This evolution is shown in Figure 1.14(a). For comparison, the marking evolution curves of places P_1 and P_2 are given for the case of a T-timed Petri net (Figure 1.11). For a VCPN, the firing speeds of the transitions are as follows:

$$\begin{cases} v_1(t) = U_1.\min(1, m_1) \\ v_2(t) = U_2.\min(1, m_2) \end{cases}$$

Note that the correspondence between the indices of the speed $v_1(t)$ of transition T_1 and the marking $m_1(t)$ of place P_1 is involuntary.

The evolution equations for the markings of places P_1 and P_2 are given in the form of differential equations:

$$\begin{cases} \dfrac{dm_1(t)}{dt} = v_2(t) - v_1(t) \\ \dfrac{dm_2(t)}{dt} = v_1(t) - v_2(t) \end{cases}$$

According to whether the value of the markings $m_1(t)$ and $m_2(t)$ is less than or greater than one, we obtain three phases of evolution: two exponential phases and one linear phase.

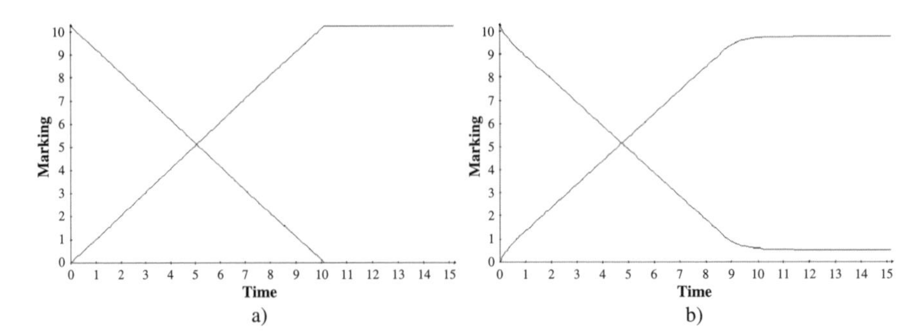

Figure 1.14. *Marking evolution curves [DAV 89]*

The evolution curves are shown in Figure 1.14(b).

The marking evolution equations for the first and second phases are given below:

– Phase 1: initial phase, $t \geq 0$, thus $m_1(t) > 1$ and $m_2(t) < 1$ as $M_0 = [10\ 0]$. As a result:

$$\begin{cases} v_1(t) = U_1 = 2 \\ v_2(t) = U_2.min(1, m_2(t)) = m_2(t) \end{cases}$$

Thus, after solving the differential equations, we obtain:

$$\begin{cases} m_1(t) = 8 + 2.e^{-t} \\ m_2(t) = 2 - 2.e^{-t} \end{cases}$$

These two (exponential) evolution equations are only valid on the time interval $[0; \ln(2)]$. Indeed, at $t = \ln(2)$, $m_2(t) = 1$. The marking evolution curve of place P_2 goes below 1. Therefore, there is a change of phase.

– Phase 2: $t \geq ln(2)$, thus $m_1(t) > 1$ and $m_2(t) \geq 1$ as $M_{ln(2)} = [9\ 1]$. As a result:

$$\begin{cases} v_1(t) = 2 \\ v_2(t) = 1 \end{cases}$$

Thus, after integrating both derivatives, we obtain two linear equations:

$$\begin{cases} m_1(t) = -t + 9 + \ln(2) \\ m_2(t) = t + 1 - \ln(2) \end{cases}$$

These two (linear) evolution equations are only valid on the time interval $[\ln(2); 8 + \ln(2)]$. Indeed, at $t = 8 + \ln(2)$, $m_1(t) = 1$. The marking evolution curve of place P_1 goes below 1. Therefore, there is a new (exponential) phase change.

REMARK 1.3.– In this example, the Petri net is bounded to 10 tokens with a closed flow of pallets. It is an invariant property given by $\forall t,\ m_1(t) + m_2(t) = 10$. This conservative property can be used to solve the system.

1.3.5. *Colored Petri nets*

When the elements being modeled are of different types, there are two possibilities: create as many Petri nets as there are types of element, or have a single Petri net with different types of tokens. Colored Petri nets provide the latter option by coloring the tokens to differentiate between them. The operation of the net remains identical to autonomous Petri nets, but with firings that depend on the colors.

DEFINITION 1.12.– *A colored Petri net is a six-tuple:*

$$(P, T, Pre, Post, M_0, C)$$

where

– P*: the set of places;*

– T*: the set of transitions;*

– $C = \{C_1, C_2, \ldots\}$*: the set of colors;*

– $Pre, Post$*: functions relating to the colors;*

– M_0*: the initial marking.*

In colored Petri nets, the tokens are assigned with different colors, which allows us to distinguish them (for example component A and B). There is also an association of functions to the arcs. Finally, places may contain tokens of different colors and the transitions have firings that depend on the colors. Colored Petri nets provide the advantage to use only one graph of Petri net and not one graph per type of token. A single net may receive different types of marks. In the case of a production machine that fabricates three types of components with different operating times, we have a single Petri net that contains three tokens of different colors instead of three Petri nets with a single type of token. The validation of the transition takes place according to a set type of color. Once the color is defined, the validation is identical to that of autonomous Petri nets:

$$M(P_i) \geq Pre(P_i, T_j/C_k), \forall\ P_i \in {}^{\circ}T_j$$

As for the validation of transitions, firing occurs according to a set color. The firing rule is identical to that of autonomous Petri nets:

$$M'(P_i) = M(P_i) + Post(P_i, T_j/C_k) - Pre(P_i, T_j/C_k)$$

An example of transition firing is shown in Figure 1.15.

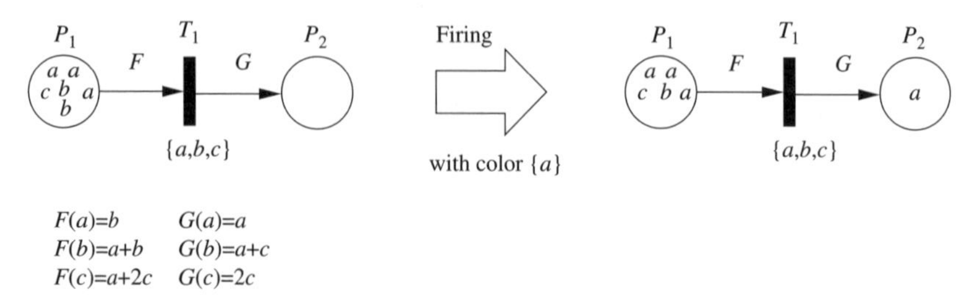

Figure 1.15. *An example of firing in a colored Petri net [DAV 89]*

EXAMPLE 1.4.– In Figure 1.15, give the markings obtained after the firing of transition T_1 according to the colors b and c. Give the marking graph of this Petri net with:

$$M_0 = \begin{bmatrix} 3a + 2b + c \\ 0 \end{bmatrix}$$

1.3.6. *Stochastic Petri nets*

Stochastic Petri nets (SPNs) were developed by Florin in 1978 to solve computational problems concerning dependability. They are regularly used to model random phenomena such as breakdowns, faults or accidents. On the basis of on Florin's initial work, numerous classes of SPNs have been created. This diversity provides a way to meet the many demands of modeling (production systems, logistic chains information systems) by integrating different types of transition (immediate, deterministic and stochastic). SPNs are undoubtedly the most widely used Petri nets in the study of production systems as it may be both a stochastic and a deterministic model at the same time.

DEFINITION 1.13.– *A stochastic Petri net is a five-tuple:*

$$(P, T, Pre, Post, M_0, \mu)$$

where

– P*: the set of places;*

– T*: the set of transitions;*

– $Pre, Post$*: forward and backward incidence mappings;*

– μ*: the set of firing rates;*

– M_0*: the initial marking.*

1.3.6.1. *Firing time*

A time d_i, which is a random variable with an exponential probability distribution, is associated with each transition T_i (Figure 1.16):

$$Pr\{d_i \leq t\} = 1 - e^{\mu_i.t}$$

$Pr\{d_i \leq t\} = Pr_{d_i}(t)$: the probability that the firing of T_i takes place before t.

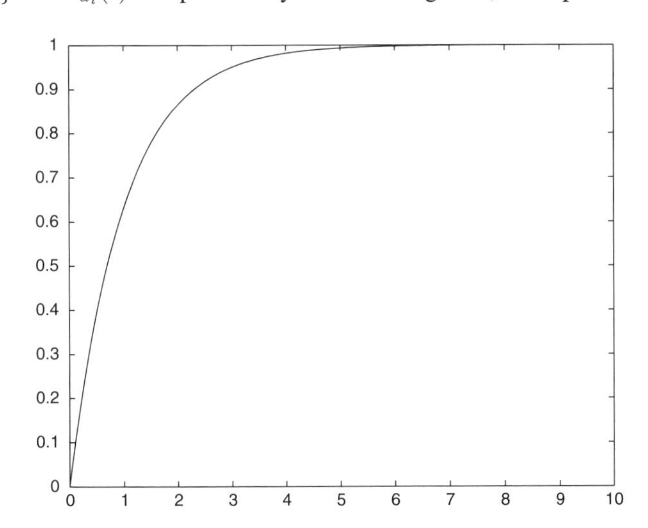

Figure 1.16. *The cumulative distribution function of an exponential distribution*

The mean value of the firing time is written as d_i^*, and is defined in the following way:

$$d_i^* = \int_0^{\infty} (1 - Pr_{d_i}(t)).dt = \int_0^{\infty} e^{-\mu_i.t}.dt = \frac{1}{\mu_i}$$

where μ_i is called the firing rate of transition T_i.

1.3.6.2. *Firing selection policy*

This policy is applied when two or more transitions are firable at the same time. In this case, it is necessary to determine which transition will be fired. Two policies are possible:

Probabilistic selection: this consists of preselecting the transition before selecting its firing time. The probability distribution is made on the subset of firable transitions. This is followed by a random selection of the transition to be fired. Once the transition is chosen, a new random selection of the firing time is made.

Random selection: the choice of transition is made after selecting the firing delays of all the firable transitions. The transition that has the smallest delay is chosen and fired with the corresponding delay. If there is an equal delay between transitions after the selection, we turn to probabilistic selection.

1.3.6.3. *Service policy*

This policy is applied when a transition is q-enabled, i.e. the transition may be simultaneously fired q times. As a result of this, three cases are possible:

– If a single selection is made, then one service at a time.

– If q selections are made, then q services at once.

– The number of selections is equal to: $Min(q, deg(T)$, where $deg(T)$ is the maximum number of simultaneous firings allowed by transition T. This condition is explicitly indicated by the transition of a loop containing $deg(T)$ tokens.

1.3.6.4. *Memory policy*

This policy is used after the firing of a transition at time t. What happens to the firing times of firable transitions? Four possibilities are available. The choice of any of the possibilities depends on the modeling of the studied system:

– Forgetting the selection (*resampling memory*): in this case, a new random selection of the firing times is carried out on the enabled transitions. This may model a breakdown phenomenon.

– Remembering the selection (*enabling memory*): the firing times of transitions remaining enabled after a decrement of t are kept. The other transitions are disabled. This may model the continuity of a phenomenon.

– Keeping the initial selection: the initial firing times (not decremented by time t) are kept, even those for transitions that have become disabled. The transitions keep this time for their next firing. This may model a job that was aborted, and then restarted from the beginning.

– Keeping the selection decremented by t: the firing times are kept, but decremented by time t, even those for disabled transitions. The transitions keep the remaining time for their next firing. This may model a job that was halted and then taken up again later.

1.3.6.5. *Petri net analysis*

The analysis of a Petri net allows its performance indices to be obtained. This analysis, which is depicted in Figure 1.17, is carried out according to the following procedure:

– construction of the marking graph;

– deduction of the associated Markov process;

– determination of the generator of the Markovian process;

– calculation of the steady-state probabilities;

– calculation of the performance indices.

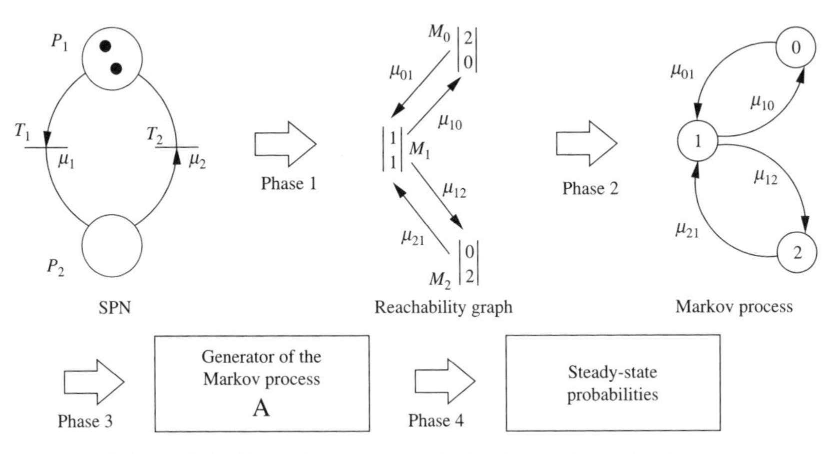

Figure 1.17. *Procedure for the analysis of a stochastic Petri net*

The different stages of this analysis are described in detail in the following sections.

1.3.6.6. *Marking graph*

This graph is obtained from M_{0+} by the successive firing of transitions. The memoryless property of the exponential distribution of transition rates shows that the marking graph of a bounded SPN is a Markov process.

1.3.6.7. *Generator of Markovian processes*

This is a square matrix denoted by A that puts together the set of transition rates among all the markings of the graph:

$$A = \begin{bmatrix} a_{11} & a_{12} & \cdots & a_{1r} \\ a_{21} & a_{22} & \cdots & a_{2r} \\ \vdots & \vdots & \vdots & \vdots \\ a_{r1} & a_{r2} & \cdots & a_{rr} \end{bmatrix}$$

where

– $A = [a_{ij}]$: an r-dimensional square matrix, where r is the number of markings;

– $a_{ij} = \mu_{ij}$: the transition rate between marking M_i and marking $M_j (i \neq j)$;

– $a_{ii} = -\sum_{j=1}^{r} a_{ij}$: the sum of the rate of exit from marking M_i.

In Figure 1.17, the generator of the Markovian process is:

$$A = \begin{bmatrix} -\mu_{01} & \mu_{01} & 0 \\ \mu_{10} & -(\mu_{10} + \mu_{12}) & \mu_{12} \\ 0 & \mu_{21} & -\mu_{21} \end{bmatrix}$$

REMARK 1.4.– Take care not to confuse the transition rates between markings with the firing rates associated with transitions. In the case where a transition T_k is q-enabled in the marking M_i and generates a new marking M_j after firing, the relation between the firing rate μ_k of transition T_k and the transition rate μ_{ij} between markings M_i and M_j is the following:

$$\mu_{ij} = q.\mu_k$$

1.3.6.8. *Fundamental equation*

This fundamental equation gives the evolution of the probability of being in a particular marking:

$$\frac{d\pi(t)}{dt} = \pi(t).A$$

where

– $\pi(t) = [\pi_1(t), \ldots, \pi_r(t)]$;

– $\pi(t) =$ the probability of being in the marking M_i at time t. This probability is also called the state probability.

1.3.6.9. *Steady-state probabilities*

The steady-state probabilities $\pi_i(\infty)$ are obtained by solving the following system of equations:

$$\begin{cases} \pi(\infty) \ . \ A = 0 \\ \sum_{i=1}^{r} \pi_i(\infty) = 1 \end{cases}$$

In effect, when $t \to \infty$, the evolution of the state probabilities $\pi_i(t)$ is constant, thus $\frac{d\pi(t)}{dt} = 0$. We find the first equation of the system. To this, we must add the fact that the sum of the steady-state probabilities is equal to 1 (as for the transient states). A system of $r+1$ equations in r unknowns (the $\pi_i(\infty)$) is thus obtained, and a solution is therefore possible.

EXAMPLE 1.5.– Consider a production machine whose evolution is determined by three distinct states: available, stopped and failed. The stochastic Petri net describing this evolution is shown in Figure 1.18. The places P_1, P_2 and P_3, respectively, model the states of availability, idleness and failure of the production machine. The firing rates associated with the transitions have the following values and meanings:

μ_i	Value	Meaning
μ_1	1	Stopping rate
μ_2	2	Restarting rate
μ_3	1	Failure rate
μ_4	3	Repair rate

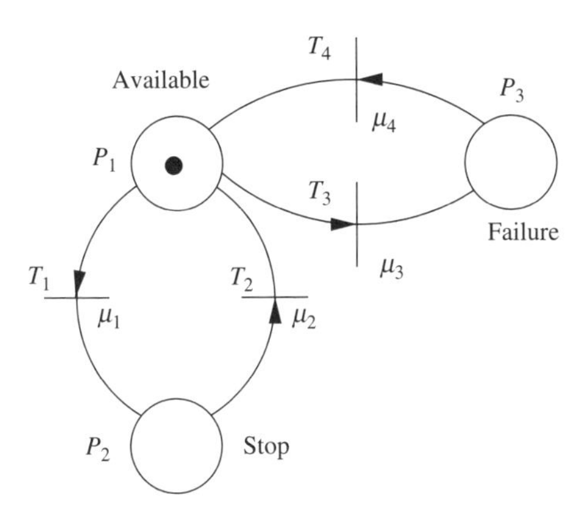

Figure 1.18. *A Petri net of the states of the production system*

From the Petri net in Figure 1.18, we obtain the graph of the Markov process. The procedure is given in Figure 1.19. We find the reachability graph composed of three markings, the Markov process associated with this graph and the evolution of the state probabilities, i.e. the evolution of the probability of being in a given marking. The generator corresponding to the Markov process is the following:

$$A = \begin{bmatrix} -\mu_1 - \mu_3 & \mu_1 & \mu_3 \\ \mu_2 & -\mu_2 & 0 \\ \mu_4 & 0 & -\mu_4 \end{bmatrix} = \begin{bmatrix} -2 & 1 & 1 \\ 2 & -2 & 0 \\ 3 & 0 & -3 \end{bmatrix}$$

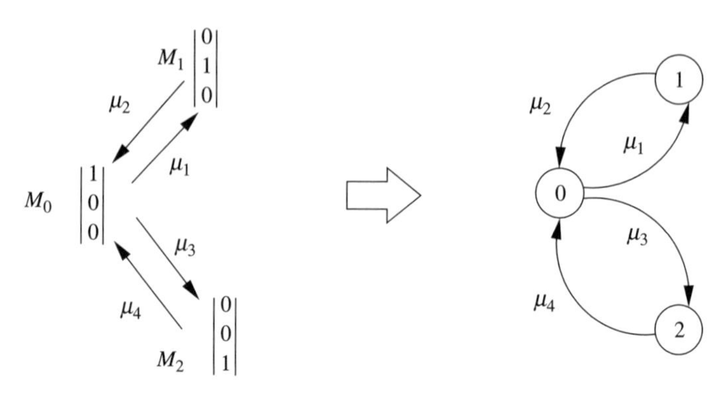

Figure 1.19. *The procedure for obtaining the Markov process*

The steady-state probabilities are calculated from the following system of equations:

$$\begin{cases} \pi(\infty) \,.\, A = 0 \\ \sum_{i=0}^{2} \pi_i(\infty) = 1 \end{cases} \Rightarrow \begin{cases} \begin{bmatrix} \pi_0(\infty) & \pi_1(\infty) & \pi_2(\infty) \end{bmatrix} \cdot \begin{bmatrix} -2 & 1 & 1 \\ 2 & -2 & 0 \\ 3 & 0 & -3 \end{bmatrix} = 0 \\ \pi_0(\infty) + \pi_1(\infty) + \pi_2(\infty) = 1 \end{cases}$$

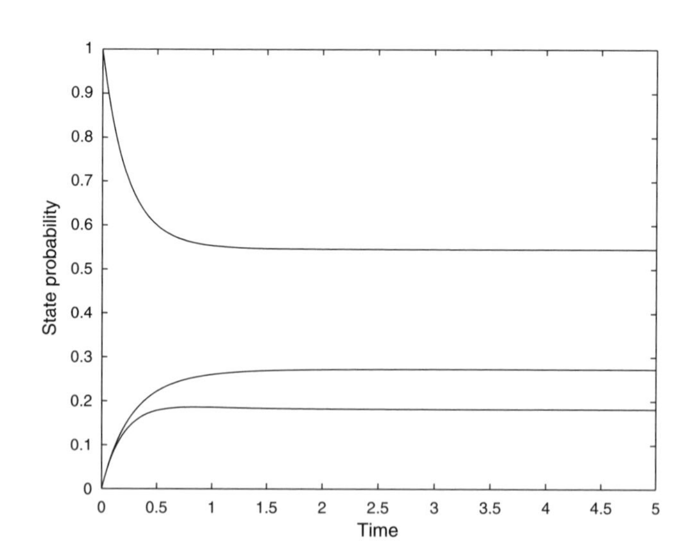

Figure 1.20. *The evolution curves of the state probabilities*

The values of the steady-state probabilities can be directly found on the evolution curve from the time $t = 2$ time units:

$$\begin{cases} \pi_0(\infty) = 0.5455 \\ \pi_1(\infty) = 0.2727 \\ \pi_2(\infty) = 0.1818 \end{cases}$$

Hence, we deduce that the machine is, on average, available 54.44%, stopped 27.27% and faulty 18.18% of the time.

1.3.6.10. *Performance indices (steady state)*

From the steady-state probabilities, it is possible to obtain different performance indices:

– the probability of a particular event E:

$$P(E) = \sum_{i \in E} \pi_i^*$$

where E is the subset of $R(M_0)$;

– the mean firing frequency of a transition:

$$f_j^* = \sum_k \mu_j(k).\pi_k^*$$

where k is such that T_j is enabled by M_k;

– the average number of tokens in a place:

$$M^*(P_i) = \sum_k M_k^*(P_i).\pi_k^*$$

– the mean time a token remains in a place:

$$D^*(P_i) = \frac{M^*(P_i)}{Post_i.F^*}$$

where

- F^*: the mean frequency of transitions upstream of P_i;

- $Post_i$: the weighting of the arcs upstream of P_i.

EXAMPLE 1.6.– Considering the example discussed previously regarding the states of a production machine, the following performance indices are obtained:

– the probability that the machine is available:

$$P(\text{machine available}) = \pi_1(\infty) + \pi_2(\infty) = 0.2727 + 0.1818 = 0.4545$$

– the mean firing frequency of transition T_1:

$$F_1^* = \mu_1(0)\pi_0(\infty) + \mu_1(1)\pi_1(\infty) + \mu_1(2)\pi_2(\infty) = \mu_1(0)\pi_0 = 0.5455$$

– the average number of tokens in place P_1:

$$\begin{aligned} M^*(P_1) &= M_0(P_1).\pi_0(\infty) + M_1(P_1).\pi_1(\infty) + M_2(P_1).\pi_2(\infty) \\ &= M_0(P_1).\pi_0(\infty) = 0.5455 \end{aligned}$$

– the mean time a token remains in place P_1:

$$D^*(P_1) = \frac{M^*(P_1)}{Post_1.F^*} = \frac{\pi_1}{[0101].[F_1^* F_2^* F_3^* F_4^*]^T} = \frac{\pi_1}{F_2^* + F_4^*} = 0.5$$

1.4. Discrete-event simulation

The simulation process consists of first designing a model for a real studied system. Experiments are then conducted with this model, which are then interpreted to make decisions regarding the real system. This procedure is often an essential tool in the evaluation of a system's performance. It calls on the principles of discrete-event simulation.

1.4.1. *The role of simulation in logistics systems analysis*

Simulation is a tool that allows us to understand the behavior of proposed or currently operating complex systems whose evolving behavior would be difficult to assess using other methods. It allows the study of the dynamic behavior of an existing system, as well as the comparison of different conceivable configurations for the design or upgradation of a system. Simulation is an important tool that makes it possible to obtain a large amount of information about a system in a very simple way.

During the design phase of a system, whether it is a mechanical system or a production system where a product will be manufactured in large quantities, it may be useful to carry out tests on the system before it has been created. The use of a simulator lets us better understand the system's operation and, for example to adjust its technical specifications or to guide the designer in the choice of the technologies to be used. In the study of inaccessible systems, simulation helps us in obtaining valuable information. We may also consider a system that cannot be decomposed into simpler, independent subsystems. If we want to study the operation of one or several subsystems independently, simulation lets us do this.

The simulation of a logistics system also becomes very useful when the creation of a mathematical model of the system is either too costly in terms of design or calculation time, or oversimplifies the reality to the point where the obtained model is no longer capable of giving useful information about the system.

Simulation is also used when observation times are incompatible with the requirements of the study to be carried out. For example, the evolution of a system may be too fast, so that on a human scale, the system's changes in state seem continuous, or too slow, such that the system does not evolve in the timescale available for the study.

Finally, note that simulation, compared to adjustments made on a real system, may be repeated several times on time frames of various lengths, and it can avoid legal or security constraints. When compared to analytic methods, it allows the study of realistic systems while taking into account complex interactions between entities, events of a random nature or any kind of probability distribution.

1.4.2. *Components and dynamic evolution of systems*

The study of systems whose operation we want to simulate begins with the modeling phase. This is followed by the simulation phase, which consists of using a computer to program and manipulate the model.

Let us consider a system with discrete events, that is a system crossed by a discrete flow, in which we may identify the moments where events leading to the evolution of the system occur. A discrete-event model is composed of *entities* (objects) with which *services* (operations) are carried out over time [FLE 06].

DEFINITION 1.14.– *An entity is defined as an individual (or a group of similar, or associated different individuals) making up the discrete flow, whose progress is followed. An entity is characterized by (fixed or variable) attributes, and may be a client, a resource, a server, a storage unit or a transport unit, for example.*

DEFINITION 1.15.– *An operation is an intentional delay in the progress of an entity. The delay may be due to the manufacturing process, transport or any other technical delay for the flow. Such services may be active or passive.*

DEFINITION 1.16.– *The dynamics of a system (or state evolution) is a description (representation) of the times at which the system changes its state due to the mechanisms that trigger and characterize the activities (services). This description consists of explaining the times at which the system changes its state. A state change occurs when an entity (machine, storage unit, etc.) changes its state.*

The states of the system are defined by the state of the objects passing through it. The state of an object is defined by the values taken by its variable attributes.

The dynamics of the system may be presented either by using the different tools outlined in the preceding sections, or by using a logical, algorithmic description, which avoids simplistic hypotheses and prepares for the writing of the simulation algorithm. To establish this logical description, we must answer the following question: which events cause the system to evolve? Once these events are identified, it is necessary to describe, in a logical manner, the evolution of the system as a function of its state when an event occurs.

1.4.3. *Representing chance and the Monte Carlo method*

In the study of logistics systems, we often encounter stochastic parameters that vary according to probability distributions, represented by random variables. To each of the values that a parameter may take, a probability is associated via a probability distribution. The study of a stochastic model requires the knowledge of the involved probability distributions to represent the system as well as the use of a tool for simulating chance to describe its operation.

What is meant by the simulation of chance? We must use a tool to decides which outcome of a random variable will actually occur. Because each outcome has an associated probability, we must use a probability generator, i.e. a random, infinite sequence uniformly distributed on $[0, 1]$.

1.4.3.1. *Uniform distribution* $U[0, 1]$

1.4.3.1.1. Definition of a pseudorandom sequence

Suppose that we have a random sequence (U_n). We may, from this sequence, obtain random outcomes of the uniform distribution $U[0.1]$. In effect, if we suppose that:

$$\forall n \in \mathbb{N}, U_n \geq 0 \qquad [1.45]$$

and we write:

$$U_{\max} = \max_{n \in \mathbb{N}} \{U_n\} \qquad [1.46]$$

then we have:

$$\frac{U_n}{U_{\max}} \in [0.1] \qquad [1.47]$$

However, in reality, there are no techniques available for the generation of a random sequence. Every obtainable sequence is *pseudorandom*. This is due to the fact that in computer science the set of different numbers that may be obtained is finite, and if we use a recurrent formula, we have:

$$U_n = f(U_{n-1}, U_{n-2}, \dots U_0) \qquad [1.48]$$

1.4.3.1.2. Obtaining a pseudorandom sequence

A pseudorandom sequence permitting the simulation of chance is better when its period (the number of terms before looping on already-obtained values) is large. Numerous generators have been developed, such as Lehmer generators (or linear congruential generators) LCG, based on a relation of the type:

$$X_n = (aX_{n-1} + b) \bmod c \qquad [1.49]$$

where a, b and c are fixed and the sequence is initialized by X_0. Thus, the sequence established, with well-chosen parameters, ensures as long a period as is possible, the greatest possible independence between successive values, and also a great ease of calculation for the machine. Algorithm 1.3 gives n outcomes of the uniform distribution $U[0, 1]$.

Algorithm 1.3 Uniform distribution algorithm

1: *Data* : a, b, c, Z_0, n
2: *Variables* : X, X_0 (real, calculation variables),R (real, result), i (integer, index)
3: $i = 0$
4: $X_0 = Z_0$
5: **while** $i < n$ **do**
6: $\quad X = (aX_0 + b) \bmod c$
7: $\quad R = X/c$
8: $\quad X_0 = X$
9: $\quad$ Write (R)
10: $\quad i = i + 1$
11: **end while**

1.4.3.2. *The Monte Carlo method*

The Monte Carlo method was developed at the end of World War II for solving partial differential equations. The name "Monte Carlo" comes from the fact that the best sequences of random numbers are given by casino roulette wheels.

This method consists of the experimental or computational simulation of mathematical or physical problems based on the selection of random numbers. We can define as a Monte Carlo method, any method aiming to calculate a numerical value using random processes, i.e. probabilistic techniques.

1.4.3.2.1. General principle

The general principle of the Monte Carlo method lies in the fact that we do not describe a phenomenon, but instead use *experimental simulations of chance*.

We must first identify the quantity m to be evaluated and a random variable X such that $E[X] = m$. Then, n outcomes of X are simulated. This is done using simulations of chance. The mean of n outcomes is then taken, which provides an estimate of m.

To obtain statistically reliable results, the simulation must be repeated a large number of times.

1.4.3.2.2. Example of application

Consider the example in Figure 1.21, supposing that the aim is to evaluate the area S.

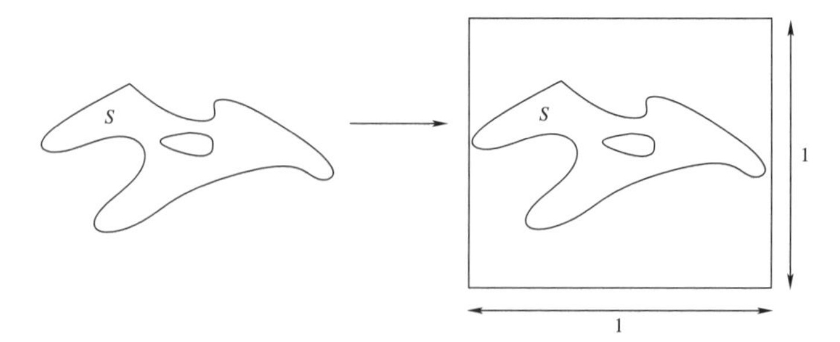

Figure 1.21. *Example application of the Monte Carlo method*

Consider the area to be enclosed in a square with sides of length 1. The quantity to be evaluated is the area S. Therefore, it is necessary to find a random variable X proportional to S. Let α be a point in the interior of the square, with coordinates (x_α,y_α) such that $0 \leq x_\alpha \leq 1$ and $0 \leq y_\alpha \leq 1$. Let X be a random variable that may take two values: 1 if $\alpha \in S$, and 0 otherwise.

Simulating n outcomes of X equates to randomly choosing n points in the square, that is randomly simulating n pairs (x_α, y_α) such that $0 \leq x_\alpha \leq 1$ and $0 \leq y_\alpha \leq 1$.

The expected value of X (which is $\sum_{i=1}^{n} X_i/n$, where X_i is the ith outcome of X) is then proportional to the ratio between the area S and the area of the square. We obtain:

$$S \approx \frac{\sum_{i=1}^{n} X_i \ \times \ \text{area of the square}}{n} \qquad [1.50]$$

Because the square has an area of 1, we obtain:

$$S \approx \frac{\sum_{i=1}^{n} X_i}{n} \qquad [1.51]$$

Algorithm 1.4 puts this principle into practice.

The Monte Carlo method is very simple to implement and allows the solution of a wide variety of stochastic and deterministic problems. It is independent of the complexity of the model and allows the visualization of the effect of different parameters and gives future orientations. On the other hand, its implementation requires a computer, and it is necessary to carry out a large number of iterations to have a high precision in terms of results.

Algorithm 1.4 Monte Carlo application algorithm

1: **Test**(X,Y) (procedure that returns 1 if the point (X_1, X_2) is in S, and 0 otherwise)
2: *Data*: N (integer, number of random selections to be made)
3: *Variables*: $X1$, X_2 (real, coordinates of points in the square), Cpt (integer, counter), i (integer, index)
4: $Cpt = 0, i = 0$
5: **for** $i = 0$ to $N - 1$ **do**
6: $\quad X_1 = U[0, 1]$
7: $\quad X_2 = U[0, 1]$
8: $\quad Cpt = Cpt + Test(X_1, X_2)$
9: **end for**
10: Write (Cpt/N)

1.4.4. *Simulating probability distributions*

Probability distributions are associated with stochastic phenomena. Here, we describe how to simulate the outcome of such a phenomenon by considering the distribution that it follows, whether it is discrete or continuous.

1.4.4.1. *Simulating random events*

1.4.4.1.1. The case of two possible events

Let X be a random variable to which we associate two outcomes A and $\overline{A}$ (see Figure 1.22).

The set of outcomes E consists of two mutually exclusive events such that:

$$P(A) = p \qquad [1.52]$$

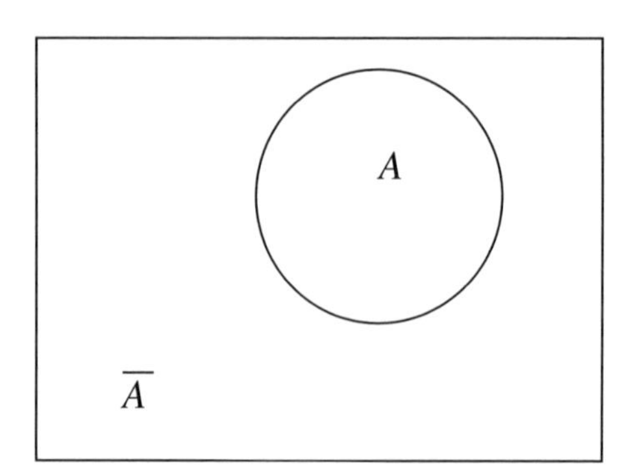

Figure 1.22. *Two mutually exclusive events*

and

$$P(\overline{A}) = 1 - p \quad [1.53]$$

To develop an algorithm that carries out a random selection giving the value taken by the random variable X, we make use of a algorithm simulating the uniform distribution $U[0, 1]$ that provides a probability, written as R. This is used to decide whether the random variable takes outcome A or $\overline{A}$. To do this, we compare R with $P(A)$ and $P(\overline{A})$ (Figures 1.23 and 1.24):

$$\text{If } R < P(A), \text{ then } X = A \quad [1.54]$$

$$\text{Otherwise, } X = \overline{A} \quad [1.55]$$

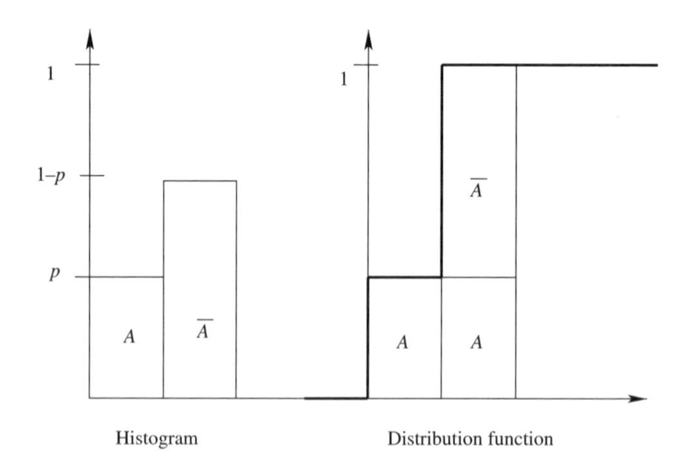

Figure 1.23. *Histogram and distribution function*

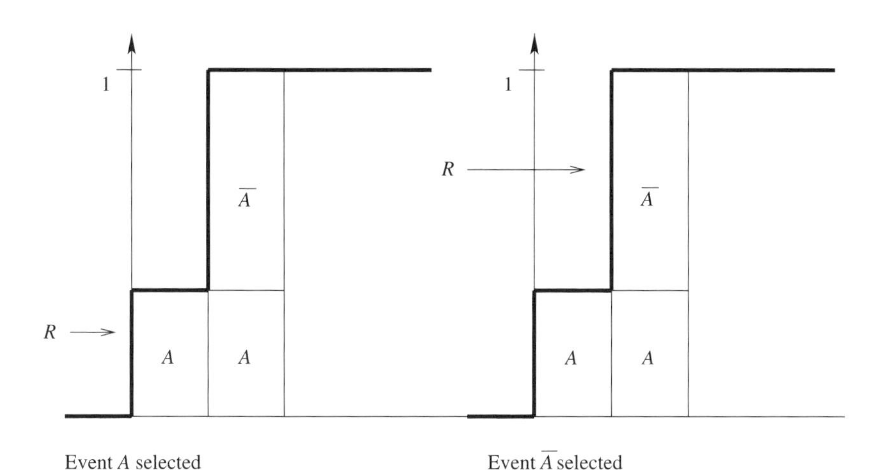

Figure 1.24. *The principle of simulating a discrete distribution*

Algorithm 1.5 illustrates this principle.

Algorithm 1.5 Simulation of two random events

1: *Data*: p (real, probability of the occurrence of event A)
2: *Variables*: R (real, variable), E (character, variable)
3: $R = U[0, 1]$
4: **if** $R \leq p$ **then**
5: $\quad E = A$
6: **else**
7: $\quad E = \overline{A}$
8: **end if**
9: Write (E)

1.4.4.1.2. Case of several mutually exclusive events

The two random events simulation approach is generalizable to the case of several mutually exclusive events. Let X be a random variable that may take n different possible values, defined by n mutually exclusive events A_i such that:

$$P(A_i) = p_i \qquad \forall i = 1, \ldots n \tag{1.56}$$

The process of simulating an outcome of the random variable X leads to the use of a simulation R of the uniform distribution $U[0, 1]$, and its comparison with the probabilities associated with the events. The events A_k $(1 \leq k \leq n)$ are chosen such that:

$$\sum_{i=1}^{k-1} p_i < R \leq \sum_{i=1}^{k} p_i \tag{1.57}$$

A particular case is that of equiprobability, where we have n events A_i such that $P(A_i) = \frac{1}{n} \quad \forall i = 1, \ldots, n$. Algorithm 1.6 allows us to simulate an outcome of a random variable X for which all the events are equiprobable.

Algorithm 1.6 Algorithm for simulating equiprobable events

1: *Data*: T (character array, such that $T[i] = A_i$), n (integer, number of events)
2: *Variables*: R (real), E (character)
3: $R = U[0, 1]$
4: $E = T[1 + \text{whole part}(nR)]$
5: Write (E)

1.4.4.1.3. Case of dependent events

Two events A and B are dependent if $P(A \cap B) \neq \emptyset$ (Figure 1.25).

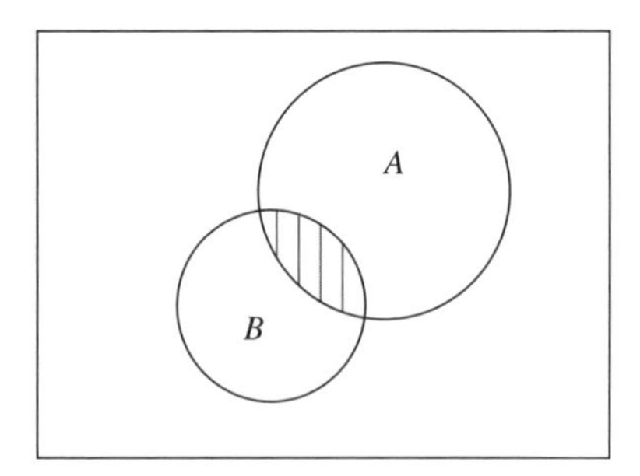

Figure 1.25. *Representation of two dependent events*

Four combinations are possible: (A, B), $(\overline{A}, B)$, $(A, \overline{B})$ and $(\overline{A}, \overline{B})$. The approach used in the simulation will consist of simulating the occurrence or non-occurrence of A, and depending on the result, simulating the occurrence of B (Algorithm 1.7).

1.4.4.2. *Simulating discrete random variables*

1.4.4.2.1. General case

Let X follow a discrete probability distribution such that $P(X = x_i) = p_i$, whose distribution function is shown in Figure 1.26.

To choose an outcome x_k of the random variable X, we use an outcome R of the uniform distribution $U[0, 1]$ and search k such that (Figure 1.27):

$$\sum_{i=1}^{k-1} P(X = x_i) < R \leq \sum_{i=1}^{k} P(X = x_i) \quad [1.58]$$

$$\Longleftrightarrow P(X \leq x_{k-1}) < R \leq P(X \leq x_k) \quad [1.59]$$

$$\Longleftrightarrow F_X(x_{k-1}) < R \leq F_X(x_k) \quad [1.60]$$

Algorithm 1.7 Algorithm for the simulation of dependent events

1: *Data*: p_A, $p_{B\backslash A}$, $p_{B\backslash \overline{A}}$ (probability distribution)
2: *Variables*: R_1, R_2 (real), E_A, E_B (characters)
3: $R_1 = U[0,1]$
4: **if** $R_1 < p_A$ **then**
5: $\quad E_A = A$
6: **else**
7: $\quad E_A = \overline{A}$
8: **end if**
9: $R_2 = U[0,1]$
10: **if** $E_A = A$ and $R_2 < p_{B\backslash A}$ **then**
11: $\quad E_B = B$
12: **else**
13: $\quad$ **if** $E_A = \overline{A}$ and $R_2 < p_{B\backslash \overline{A}}$ **then**
14: $\quad\quad E_B = B$
15: $\quad$ **else**
16: $\quad\quad E_B = \overline{B}$
17: $\quad$ **end if**
18: **end if**
19: Write (E_A, E_B)

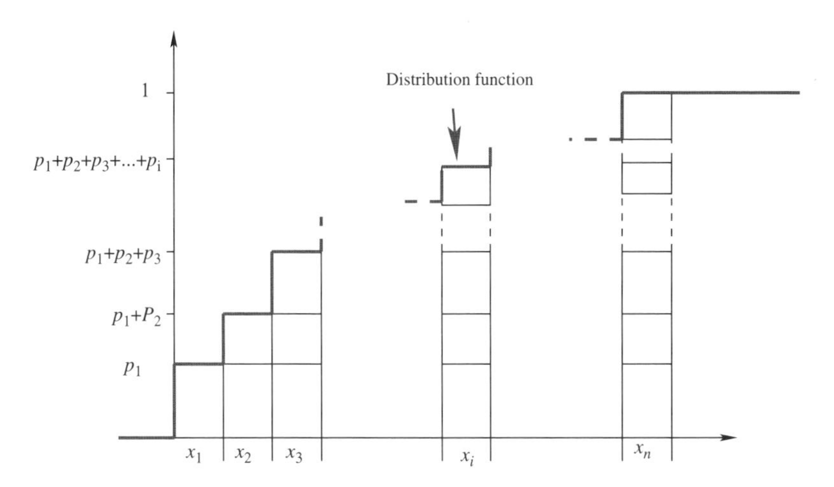

Figure 1.26. *The distribution function of a discrete law*

Depending on the distribution function $F_X(x)$, we may simplify the above equation and develop algorithms for each probability distribution.

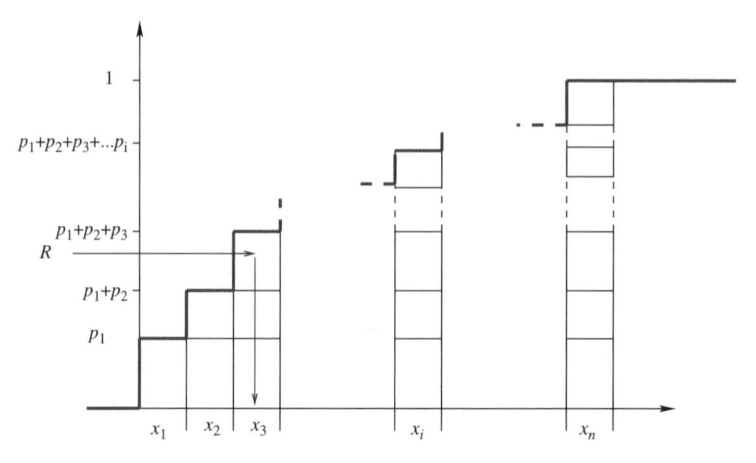

Figure 1.27. *The random generation of an outcome in the case of a discrete distribution*

Let us take the example of a random variable X that follows the Poisson distribution with parameter λ. It can take the values $0, 1, 2, \ldots$ such that:

$$P(X = k) = \frac{\lambda^k}{k!} \exp(-\lambda) \quad k \in \mathbb{N} \qquad [1.61]$$

Note that the support of the distribution is infinite, and it is defined for $k \geq 0$. It is straightforward to show that:

$$P(X = k + 1) = \frac{\lambda P(X = k)}{k + 1} \qquad [1.62]$$

and obtains the simulation algorithm (Algorithm 1.8).

Algorithm 1.8 Poisson distribution simulation algorithm

1: *Data*: λ (parameter of the probability distribution)
2: *Variables*: $R,\ S,\ var$ (real), k (integer)
3: $k = 0$, $R_1 = U[0, 1]$
4: $S = \exp(-\lambda)$
5: $var = \exp(-\lambda)$
6: **while** $S < R$ **do**
7: $\quad k = k + 1$
8: $\quad var = var * \lambda/k$
9: $\quad S = S + var$
10: **end while**
11: Write (k)

1.4.4.3. *Simulating continuous random variables*

In the case of a discrete distribution, the approach consists of using a simulation of the distribution $U(0,1)$ and comparing it with the distribution function. In the case of a continuous probability distribution, we use the same general approach, but with a continuous distribution function and its inverse function.

1.4.4.3.1. Inverse function method

For a continuous probability distribution, we obtain the distribution function $F_X(x) = P(X \leq x)$ from the probability density $f(x)$ using the following formula:

$$F_X(x) = \int_{-\infty}^{x} f(t)dt \qquad [1.63]$$

This function is defined on the set of possible outcomes of X and takes its values in the interval $[0,1]$. It is strictly increasing, continuous, bijective and, therefore, has an inverse function F^{-1}. This inverse function is defined on $[0,1]$ and takes its values on the set of outcomes of X (Figure 1.28). We randomly select an outcome of the distribution $U[0,1]$ and obtain a corresponding outcome such that:

$$X_k = F^{-1}(R) \qquad [1.64]$$

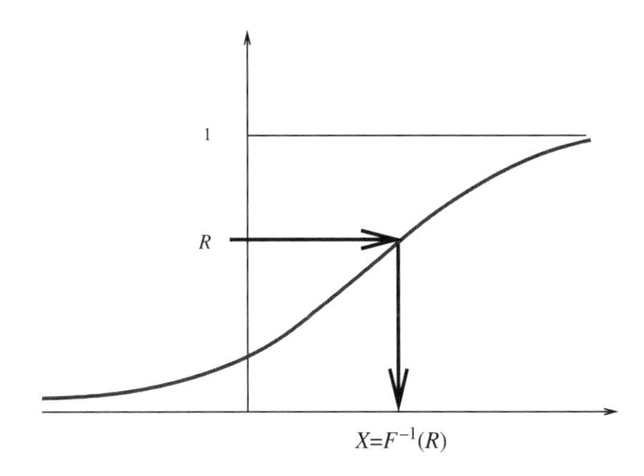

Figure 1.28. *The inverse function method*

This inverse function approach can be applied to numerous functions, but there are some cases such as the normal distribution, for which it does not work.

1.4.4.3.2. Case of the normal distribution

A random variable X follows the normal distribution $N(m,\sigma)$ if the probability density associated with an outcome x is given by the following function:

$$f(x) = \frac{1}{\sigma\sqrt{2\pi}} \exp \frac{-(x-m)^2}{2\sigma^2} \qquad [1.65]$$

This function is a probability density function, but we cannot explicitly write the corresponding distribution function.

One method relies on the central limit theorem to simulate an outcome of the normal distribution $N(0,1)$, known as the standard normal distribution. The central limit theorem considers a sequence $X_1, X_2, \ldots, X_n$ of independent and identically distributed random variables, with mean value μ and variance σ^2. Then, the distribution of:

$$\frac{X_1 + X_2 + \ldots + X_n - n\mu}{\sigma\sqrt{n}}$$

tends to a standard normal distribution as n tends to infinity.

On the basis of this principle, we simulate a large number n of outcomes of R_i ($i = 1, \ldots, n$) of the distribution $U[0,1]$. We use them to establish an outcome of the normal distribution $N(0,1)$. Suppose that the mean of these n outcomes is $\frac{1}{2}$ and the standard deviation is $\frac{1}{\sqrt{12}}$.

Thus, we have:

$$\frac{\sqrt{12}}{\sqrt{n}}(R_1 + R_2 + ... + R_n - \frac{n}{2}) \rightarrow \mathrm{N}(0,1) \qquad [1.66]$$

Once an outcome R of the standard normal distribution has been obtained, the outcome of any normal distribution $N(m,\sigma)$ can be obtained by a simple change of variable:

$$X = \sigma R + m \qquad [1.67]$$

Also there exists the Box–Muller method, which takes two outcomes R_1 and R_2 of the distribution $U(0,1)$ and obtains two outcomes X_1 and X_2 of the distribution $N(0,1)$, using the following formulas:

$$X_1 = \sqrt{-2ln(R_1)}cos(2\pi R_2) \qquad [1.68]$$

$$X_2 = \sqrt{-2ln(R_1)}sin(2\pi R_2) \qquad [1.69]$$

1.4.4.3.3. Complex continuous random variables

The inverse function method cannot be applied when F^{-1}, the inverse distribution function, cannot be explicitly stated. We then make use of other approaches to design simulation algorithms.

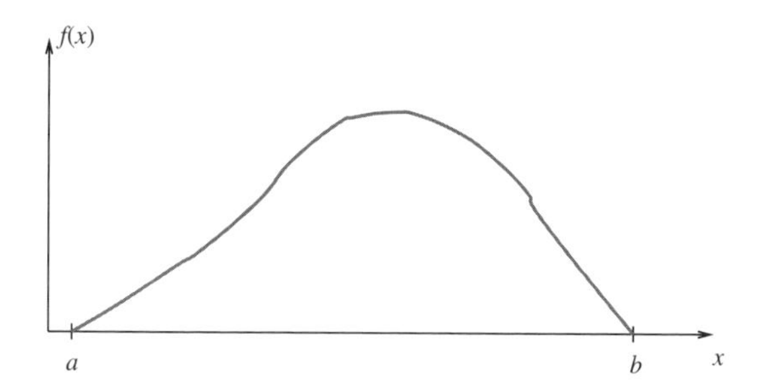

Figure 1.29. *The approximation method: probability density*

1.4.4.3.4. The approximation method

This method is based on the decomposition of the interval of definition, written $[a, b]$, of the random variable X (with probability density $f(x)$) for which we want to simulate an outcome (Figure 1.29).

The interval $[a, b]$ is decomposed into n intervals $[a_i, a_{i+1}]$ ($i = 0, \ldots, n-1$), which are not necessarily of the same width, but of the same probability (equation [1.70]):

$$P(a_i < X < a_{i+1}) = \frac{1}{n} \quad [1.70]$$

$$\Longleftrightarrow \int_{a_i}^{a_{i+1}} f(x)dx = \frac{1}{n} \quad [1.71]$$

$$\Longleftrightarrow F(a_{i+1}) - F(a_i) = \frac{1}{n} \quad [1.72]$$

$$\Longleftrightarrow F(a_{i+1}) = \frac{1}{n} + F(a_i) \quad [1.73]$$

Because the function F^{-1} does not exist, we must therefore make an approximation. $F(a_{i+1}) - F(a_i)$ represents the area of the surface contained between the segment $[a_i, a_{i+1}]$ and the curve of $f(x)$ (Figure 1.30).

This area is approximated by that of the rectangle of width $[a_i, a_{i+1}]$ and height $f(a_i)$. Thus, we obtain:

$$\int_{a_i}^{a_{i+1}} f(x)dx = \frac{1}{n} \quad [1.74]$$

$$\Longleftrightarrow f(a_i)(a_{i+1} - a_i) \approx \frac{1}{n} \quad [1.75]$$

$$\Longleftrightarrow a_{i+1} \approx \frac{1}{nf(a_i)} + a_i \quad [1.76]$$

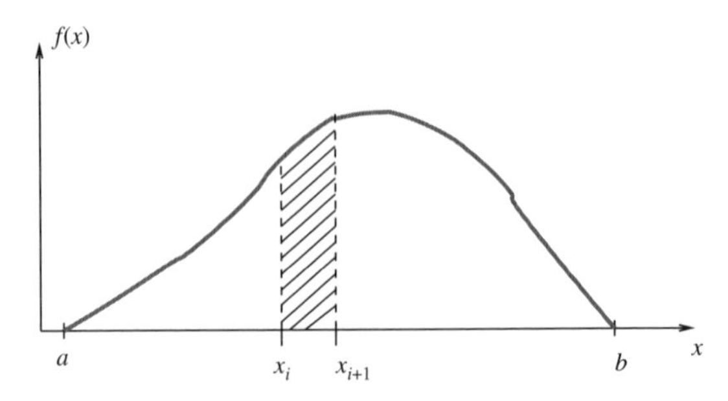

Figure 1.30. *The approximation method: definition of the interval* $[a_i, a_{i+1}]$

Therefore, the approximation method will be described in the following way:

– separate the interval $[a, b]$ into n intervals of the same probability;

– randomly select an interval $[a_i, a_{i+1}]$ among the n intervals (simulation of an equiprobable discrete distribution);

– simulate an outcome in the selected interval using a uniform distribution $U[a_i, a_{i+1}]$.

1.4.4.3.5. The exclusion method

Let X be a random variable whose probability density is defined on an interval $[a, b]$ (Figure 1.31). The value $f_{\max}$ is the maximum value of this function on its interval of definition.

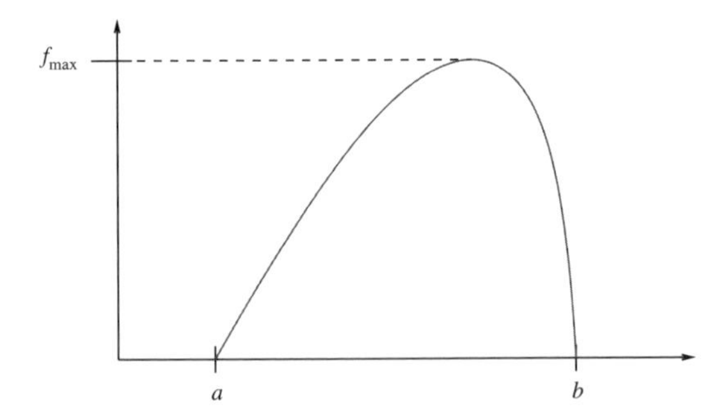

Figure 1.31. *The exclusion method: probability density*

Consider, on the one hand, an outcome R of the uniform distribution $U[a, b]$ (Figure 1.32) and its image under f, written as $f(R)$, and on the other hand, an outcome R' of the uniform distribution $U[0, f_{\max}]$. The value given by R is considered as an outcome of X if we have $R' < f(R)$ (Figure 1.33). In the case where this condition is not satisfied, it is necessary to randomly select another pair (R, R') until the condition is satisfied.

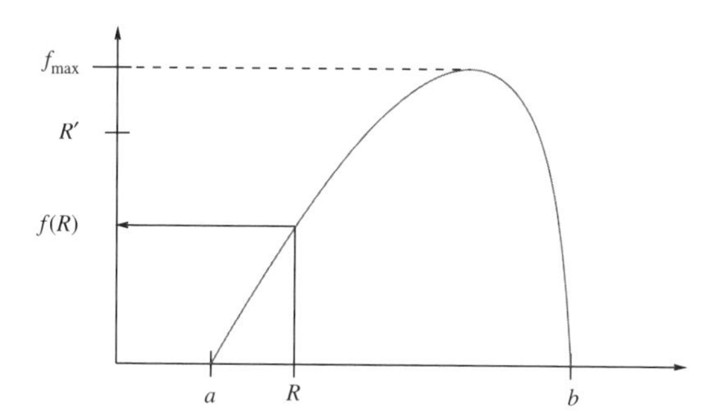

Figure 1.32. *The exclusion method (1)*

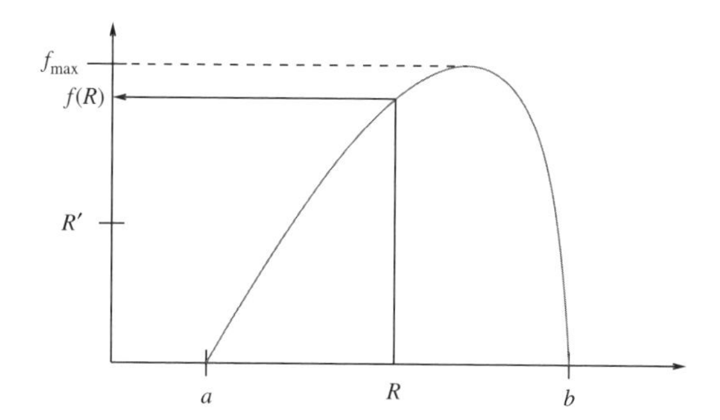

Figure 1.33. *The exclusion method (2)*

1.4.4.3.6. The superposition method

The superposition method is based on the distribution function $F_X(x)$. It is applied in the case where the distribution function may be written in the following way:

$$F_X(x) = \sum_{i=1}^{n} p_i F_i(x) \quad [1.77]$$

where

$$\sum_{i=1}^{n} p_i = 1 \quad [1.78]$$

Each function F_i is invertible such that $r = F_i(x) \Longleftrightarrow x = F_i^{-1}(r)$.

The principle behind the superposition method is therefore the following:

– randomly select (with the algorithm for a discrete distribution) an integer k between 1 and n where a weight p_i is associated with each integer $1 \leq i \leq n$;

– $x_k = F_k^{-1}$(R), where R is an outcome of $U[0, 1]$, is an outcome of the initial random variable X.

1.4.5. *Discrete-event systems*

The objective of the simulation of discrete-event systems is to reproduce, event-by-event, the time evolution of a system based on an established model. It is necessary to reproduce for each entity in the model, the state changes of the corresponding entity in the real system.

The models considered here are for discrete-event dynamic systems (DEDS). Changes in state occur instantaneously at discrete moments in time.

We define a system based on the nature of the flow running through it (Table 1.3). In the case of discrete flow, it is made-up of separable, countable and individually identifiable entities (for example clients and mechanical parts). We define an event as being the instant of the change of state of a system. In the case of continuous flow, we cannot isolate something in the flow, and what happens to the flow is expressed mathematically by rates of change over time (for example the flow of heat, liquids and gasses). Therefore, time management is not done in the same way for both types of systems. In the continuous case, every Δt time units, the flow and system state variables are updated. In the discrete case, time management is not based on a concern for regularity. A discrete model progresses in time by constantly asking the following question, “at what time will the next change in the system occur?”

	Continuous systems	Discrete systems
Entities	Inseparable	Separable
Evolution of the flow	Mathematical expressions	Events
Evaluation	By calculation	By counting
Time management	T_{start}, T_{end}, Δ_t between two measurements	Irregular jumps

Table 1.3. *The main differences between discrete-flow and continuous-flow systems*

1.4.5.1. *Key aspects of simulation*

To carry out a simulation of a discrete-event system, we must use basic “building blocks” and “automatic mechanisms” that allow the description of the dynamic operation of the system, i.e. its state change logic. These building blocks and mechanisms must make random selections from given probability distributions, automatically advance the time (clock), progressively record the statistics necessary for analysis and develop a simulation report.

1.4.5.1.1. Description of the system

In general, the considerations presented above are adopted. We consider a flow of entities crossing the system. Each of the entities is characterized by some (fixed or variable) attributes. According to certain conditions, the entities interact with the resources and servers, as well as with global variables. These interactions correspond to events, and may also generate events.

Once we have all the structural data necessary for the description of the system, we move on to the description of the state change logic of the system, i.e. its dynamics.

The states of the system correspond to the values of the variable attributes of the entities and resources, the states of queues and the values of global variables. The occurrence of events may cause state changes. Indeed, an event initializes an activity and leads to a state change in the objects involved in this activity. The state of the system remains constant between two successive events.

1.4.5.1.2. Procedure

The procedure for describing the system may be summarized in three stages:

– make a graphical representation of the flow in the system;

– identify the entities and their attributes;

– establish the dynamics by asking the following question: which events modify the state of the system? From this question, we describe the dynamics by events. For each event, we identify the state of the system and determine its future (i.e. that of the entity and the server).

The different events that occur correspond to the state changes of the system. The different types of events may be grouped into four categories:

– the arrival of a client in the system;

– the exit of a client from the system;

– the beginning of a client's processing;

– the end of a client's treatment.

First, it is necessary to find the types of events that generate changes in the system and group them into categories. An event is represented by:

– a date;

– an event type;

– a reference to the involved active entity.

The method of management by event is based on a scheduler module whose role is to arrange the initial events into a timeline, arrange the events created during the evolution of the model, activate the treatments associated with events in terms of their date of occurrence and update the current date of the simulation, as well as manage the termination of the simulation.

1.4.5.1.3. Example of an application

In a television (TV) set assembly process, there is a post where the quality control of TVs leaving the production line is carried out. At this post, two controllers work in parallel in a completely identical manner. When a TV does not work as it should, it is discarded. When a TV set meets the requirements for satisfactory control, it is sent to be packed. The size of the buffer, or the waiting zone, which specifies the amount of the stock-in-progress arriving at the control post, is *a priori* infinite. We suppose that the TVs arrive, on average, every m minutes (exponentially distributed) from the production line, and that a controller carries out their operations in $t \pm \delta$ minutes (uniformly distributed).

The corresponding flow diagram is shown in Figure 1.34.

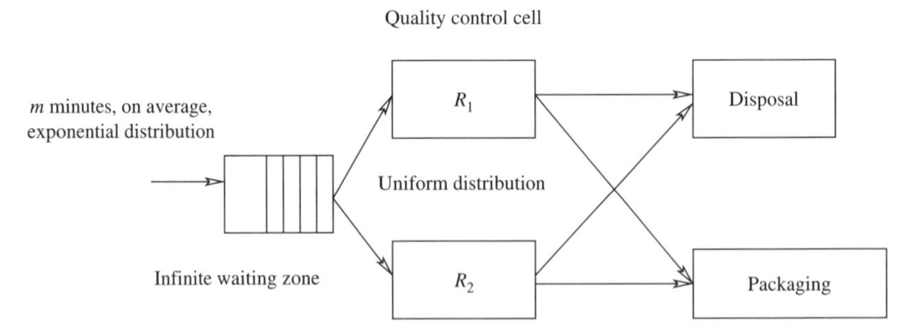

Figure 1.34. *Flow in the quality control system*

Then, we define the different objects:

– clients (TVs):

- fixed attributes: the interarrival time follows an exponential distribution with parameter $(1/m)$;

- variable attributes: these are the possible states of the TVs in the system, i.e. awaiting control, being assessed, in good condition and in poor condition.

– the servers (the two control posts):

- fixed attributes: the processing time between successive TVs: $U[t-\delta, t+\delta]$;

- variable attributes: this is the state of the server: free or occupied.

– the storage unit at the input of the control cell:

- fixed attributes: its capacity: infinite;

- variable attributes: the number of TVs inside.

We then move on to the identification of the services:

– active service: quality control: $U[t-\delta, t+\delta]$;

– passive service: waiting in the input stock.

To be able to explicitly describe the system dynamics, we make certain hypotheses: when a TV arrives and if both controllers are available, we can put a priority rule in place, for example, and say that R_1 will take the set, or we may imagine that a random selection is made corresponding to a draw. Here, we are not concerned with a TV's probability of being defective, and therefore the probability that it is discarded, but we may include it and say that a TV is discarded with a probability of p, and is therefore sent to be packaged with a probability of $(1-p)$. To explicitly describe the system's dynamics, it is necessary to ask the following question: which events change the state of the system? Here, we have two events: the arrival of a set in the system and the end of its processing by a controller.

Generally, when a client arriving from outside the system is concerned, we must carry out tests on the state of the system to know is future. When the server has finished processing a client, we must ask two questions: what is next for the client and what is next for the server. In our case, this amounts to writing:

– when a TV arrives in the system:

- If R_1 or R_2 are free (that is the queue is empty), the TV goes to both of them (depending on the chosen priority rule and/or their availability). Generate a processing time according to the distribution $U[t-\delta, t+\delta]$.

- If R_1 and R_2 are busy, the TV goes into the queue.

- Generate the date of the next arrival of a TV (according to a $(1/m)$ exponential distribution).

– when R_1 or R_2 has finished processing a TV:

- If the TV works, it goes to be packaged; otherwise, it is discarded.

- If the queue is not empty, a TV is taken from it and we generate the processing time according to the distribution $U[t-\delta, t+\delta]$; otherwise, the controller waits.

1.4.5.1.4. Integrating data to carry out a random selection

Once the model has been defined, the quantitative data such as the operation time, breakdown time and rejection rate, etc. are integrated. These data may be deterministic or stochastic. Sometimes existing data are used, and at other times it is necessary to collect information to determine the distribution that an identified random variable follows.

We must avoid simply reading an existing record, as we will only be able to reproduce previously observed behavior. Thus, it is necessary to focus on a selection from an empirical distribution (i.e. a histogram) or a selection using a standard theoretical distribution.

The principal probability distributions that are encountered are the Poisson distribution (which represents a number of arrivals or services during a period of time), the uniform distribution $U[a, b]$ (used in the case where all of the real values between a and b have the same probability), the exponential distribution (which often represents interarrival times or service times with a server that does not become worn out) and the Gaussian or normal distribution (used to characterize fluctuations around a mean value, brought about by multiple causes).

1.4.5.1.5. Time management

In the simulation of a discrete-event system, the state changes in the system occur at irregular intervals. We will therefore advance in time by making jumps.

To do this, we must first specify the initial state of the system, that is the initial values of the variable attributes of objects, such as, the state of the machines at time $t = 0$. Likewise, we must specify the conditions under which the simulation terminates, for example when a simulation time frame H is reached, when x parts are produced and when a stock is sufficiently replenished.

To manage the evolution of time, the use of a clock, incrementing per fixed unit of time and searching for events that occur at each increment, is not suitable. It is preferable to use event-based management.

Therefore, we have a schedule in which the moments of occurrence of the next events whose date is known are recorded. Time is incremented from one date to another, the order being modified by the treatment of events that generate further events.

Algorithm 1.9 Event-scheduling algorithm

```
1: Variables : t (real): clock, t_ev (real): occurrence date of an event
2: t=0
3: while the termination condition is not satisfied do
4:    take the first event in the list, t = t_ev
5:    delete this event from the list
6:    call the procedure of the deleted event
7:    if the procedure has created new events then
8:       place them in the list
9:    end if
10: end while
```

1.4.5.1.6. Transitional regime

In the case of stochastic models, we encounter transitional regimes for certain variables, such as a level of stock. This is due to the fact, among others, that the behavior of the model depends on the random sequence that generates events and makes the evaluated parameter evolve.

Recording information about a transitional regime will not reveal the behavior of the system; the information must be gathered in the steady (or stationary) state. For this, it is possible, for example to include initial conditions that resemble the steady state, to conduct simulations for long enough time to render the transitional regime insignificant or to remove the values recorded during the transitional regime from the final statistics.

1.5. Decomposition method

1.5.1. *Presentation*

The decomposition method, developed by Gershwin [GER 87] in 1987, is another tool for the performance evaluation of production lines. Its basis consists of dividing a system into several smaller subsystems. Let us take the example of the line (L) in Figure 1.35 with M workstations. Between two workstations (or machines) (m_i and m_j), we make use of a buffer stock, written b_k. Therefore, there exist $M - 1$ buffers in the line.

This line may be decomposed into $M - 1$ systems with two stations and a buffer. More precisely, we consider a set of lines with two pseudo-machines. Each line has a buffer b_i, an upstream station ($M_u(i)$) and a downstream station ($M_d(i)$).

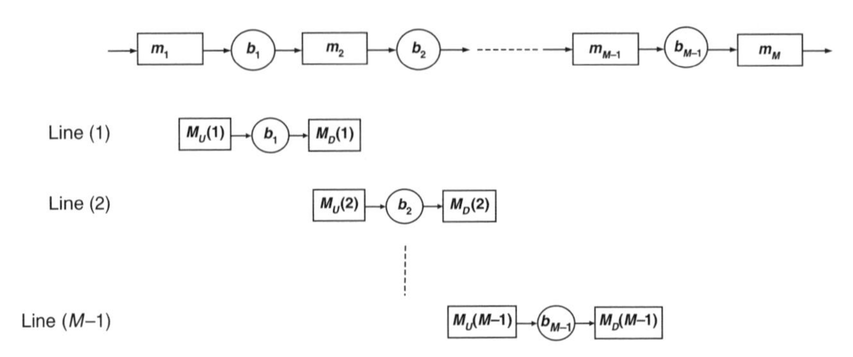

Figure 1.35. *Structure of the decomposition method [GER 94]*

The parameters of these lines are determined by relations between the flow of products crossing the buffers of the original system. These relations are called the flow conservation equations.

The main advantage of this method is that it efficiently estimates the performance of a line during an execution time, which is relatively short in comparison with simulation models [GER 94]. This advantage was tested following a comparison between two tools: the decomposition method and simulation with the ARENA® software. The comparison was made on an assembly line with eight stations and seven buffers.

We noted that the decomposition method estimates the performance of the line with only a 3% difference from the simulation method. Hence, the advantage is the significant reduction in calculation time, which is, on average, less than a second for the decomposition method, compared with 5 m for simulation with ARENA®. However, this method does not allow the consideration of the transitional regime of the studied system, which may be done with discrete-event simulation. Therefore, with the decomposition method, we cannot examine the evolution of the system in the different states that it may exist.

1.5.2. *Details of the method*

Let us again consider the line in Figure 1.35, with finite-sized buffers and stations with machines whose mean time between failures and mean repair time follow an exponential distribution.

The main line L with M stations and $M-1$ buffers is decomposed into a set of two-station lines $L(i)$ $(i = 1, \ldots, M-1)$. Their buffers (b_i) have the same sizes as those of L.

Each pseudo-machine $M_u(i)$ models the part of the line upstream of buffer b_i, while $M_d(im)$ models the downstream part of b_i.

Theoretically, the parameters of the pseudo-machines are chosen as follows:

– The flow rate entering and leaving buffer B_i in line $L(i)$ is approximately equal to that of buffer b_i in line L.

– The probability that line $L(i)$'s buffer is empty or full is close to that of buffer b_i in line L.

– The probability of resuming the flow across line $L(i)$'s buffer in an interval of time after a period during which this flow was interrupted is close to the probability of the corresponding event in line L.

– The average number of products in a buffer in line $L(i)$ is approximately the number in buffer b_i of line L.

To calculate the values of the pseudo-machines' parameters, the system is modeled in the form of a Markov process whose steady-state operational behavior is studied by considering sublines.

The parameters of the pseudo-machines are $r_u(i)$, $p_u(i)$ and $\mu_u(i)$ for the first machine and $r_d(i)$, $p_d(i)$ and $\mu_d(i)$ for the second machine. Other parameters are taken into account:

– $n_i(t)$: the number of items in buffer b_i at time t;

– $r_u(i)\delta t$: the probability that machine i goes from a state of failure to its functional state during the time interval (t, $t + \delta t$) such that δt is negligible;

– $p_u(i)\delta t$: the probability that machine i goes from its functional state to a state of failure during the time interval (t,$t + \delta t$), given that 0 < n_i < N and δt is negligible;

– $\mu_u(i)\delta t$: the number of items treated by machine $M_u(i)$ during the time interval $(t, t + \delta t)$, with δt negligible;

– $r_d(i)\delta t$: the probability that machine $M_d(i)$ goes from a state of failure to its functioning state during the time interval (t, $t + \delta t$), with δt negligible;

– $p_d(i)\delta t$: the probability that machine $M_d(i)$ goes from its functioning state to a state of failure during the time interval (t, $t + \delta t$), given that $0 < n_i < N$ and δt is negligible;

– $\mu_d(i)\delta t$: the number of items treated by machine $M_d(i)$ during the time interval $(t, t + \delta t)$, with δt negligible.

The state of the Markov process defined by Gershwin [GER 87], which represents the states of the system $L(i)$, is defined as follows:

– s_i: $(n_i, \alpha_u(i), \alpha_d(i))$;

– $0 \leq n_i \leq N$: number of items in buffer i;

– $\alpha_u(i)$: the states of pseudo-machine $M_u(i)$ (0: not operational and 1: operational);

– $\alpha_d(i)$: the states of pseudo-machine $M_d(i)$ (0: not operational and 1: operational).

The distribution of states of system s_i is calculated based on the function $f_i(n_i, \alpha_u(i), \alpha_d(i))$ as well as the probabilities of the events $p_i(0, 0, 1)$; p_i(0,1,1); p_i(N, 0, 1) and p_i(N, 1, 1), as equations [1.79]–[1.82] show (see [BUR 97] for detailed calculations):

$$r_u(i)\delta t = prob(\alpha_u(i) = 1 \text{ to } t + \delta t \ll \alpha_u(i) = 1 \text{ to } t) \quad [1.79]$$

$$r_d(i)\delta t = prob(\alpha_d(i) = 1 \text{ to } t + \delta t \ll \alpha_d(i) = 0 \text{ to } t) \quad [1.80]$$

$$p_u(i)\delta t = prob(\alpha_u(i) = 0 \text{ to } t + \delta t \ll \alpha_u(i) = 1 \text{ to } t) \quad [1.81]$$

$$p_d(i)\delta t = prob(\alpha_d(i) = 1 \text{ to } t + \delta t \ll \alpha_d(i) = 0 \text{ to } t) \quad [1.82]$$

Finally, the efficiencies of machines $M_u(i)$ and $M_d(i)$ are defined as shown in equations [1.83] and [1.84]. These efficiencies allow the determination of the line's output rate based on the efficiency of machine $M_d(M - 1)$:

$$E_u(i) = prob(\alpha_u(i) = 1, n_i(t) < N_i) + p_i(N, 1, 1).(\frac{\mu_d(i)}{\mu_u(i)}) \quad [1.83]$$

$$E_d(i) = prob(\alpha_d(i) = 1, n_i(t) > 0) + p_i(0, 1, 1).(\frac{\mu_u(i)}{\mu_d(i)}) \quad [1.84]$$

Chapter 2

Optimization

2.1. Introduction

Once we have a tool to assess the behavior and the performance of a studied system, we seek to optimize its functioning. This may be a single-criterion study, such as the minimization of cost or duration, or a multi-objective study in which we seek the best compromise between different performance indicators such as the joint minimization of the cost associated with a set of vehicle routes and the number of vehicles used. The optimization problems that are encountered while analyzing logistics systems are often combinatorial, and are characterized by the fact that as the number of possible solutions (or combinations) increases, the size of the problem considerably (exponentially) increases. Consequently, the computation time required for the enumeration of all combinations is directly affected. A large number of these optimization problems are NP-hard [GAR 79]. This means that if we wish to find the optimal solution to such a problem, it is necessary to list all the possible solutions. This enumeration requires a computation time which increases exponentially with the size of the problem. Hence, the decision support tool used to solve logistic system problems must give an answer quickly, and the best possible answer. This leads us to classify optimization methods into different categories: exact optimization methods, which are costly in computation time but guarantee the optimality of the solution, and approximate (heuristic and metaheuristic) methods, which are faster, but do not guarantee the optimality of the obtained solution.

The first step to be taken before developing or applying an optimization method is to model the problem mathematically (mathematical programming). Then, we must consider the size of the problems that are encountered in the industrial world and the decision level of the problems in question. It is reasonable to allow greater computation times for the resolution of tactical or strategic problems than for an operational management or real-time problem.

2.2. Polynomial problems and NP-hard problems

2.2.1. *The complexity of an algorithm*

The consideration of complexity consists of, under a certain number of conditions, evaluating the computation time and memory space required to arrive at the result of an algorithm [GAL 10]. Controlled complexity is synonymous with a predictable time and a bounded memory space.

In the space of 50 years, the computational capacity of machines has increased from less than a million floating-point operations per second to more than a million billion floating-point operations per second. This evolution calls for a fundamental change in the development and handling of algorithms, in particular their implementation. Computational needs are important. It suffices to note that ordering a list of size n requires $n!$ operations. It is then easy to see that limits are quickly reached in the case of direct use. As an example, if we consider that for each configuration a microsecond is necessary per operation, then a list of 18 positions will require more than 205 years of continuous calculation (Table 2.1).

Size	Number of possibilities $n!$	Time (1 microsec./operation)	Days	Months	Years
1	1	0.000001	-	-	-
2	2	0.000002	-	-	-
3	6	0.000006	-	-	-
4	24	0.000024	-	-	-
5	120	0.00012	-	-	-
6	720	0.00072	-	-	-
7	5,040	0.00504	-	-	-
8	40,320	0.04032	-	-	-
9	362,880	0.36288	-	-	-
10	3,628,800	3.6288	-	-	-
11	39,916,800	39.9168	-	-	-
12	479,001,600	479.0016	-	-	-
13	6,227,020,800	6,227.0208	-	-	-
14	87,178,291,200	87,178.2912	1.009008	-	-
15	1.30767E+12	1,307,674.368	15.13512	-	-
16	2.09228E+13	20,922,789.89	242.16192	8.072064	-
17	3.55687E+14	355,687,428.1	4,116.75264	137.225088	11.435424
18	6.40237E+15	6,402,373,706	74,101.54752	2,470.051584	205.837632

Table 2.1. *The computation time as a function of the complexity of the studied problem*

We may then ask whether a partial exploration will be sufficient. The problem is then treated in an approximate and intelligent way. There is another computational limit of a physical type and it is related to the computer's performance. Galtier and Laugier [GAL 10] note that "computational power will grow, the cost per floating-point operation tends to stagnate. The total global energy consumed by computers was 200 TWh in 2005, which corresponds to the power from around 30 nuclear reactors, each running at 800 MW, or in terms of their carbon footprint, 30 million automobiles. The most powerful computer in the world in 2009, the Cray Jaguar, consumes 7 MW for a computational power of 1.759 PFlops. This cost, requested by the TOP500 organization, is only a partial cost of calculation. High-performance calculation specialists estimate that very soon the total energy costs of a supercomputer will be in the order of 100 MW."

2.2.2. *Example of calculating the complexity of an algorithm*

Let us take again the previous example concerning the treatment of ordering a list. If we seek to create this ordered list by following a given classification rule, this is known as an order or sorting algorithm. This type of algorithm is among the most popular and widely used algorithms. We will analyze the complexity of some sorting processes to illustrate the impact of algorithm on computation times.

One of the most well-known methods consists of going through the whole list, selecting the element with the greatest value, removing it from the current list and then inserting it at the top of the list to be sorted. Since this method involves going through the whole list, it necessitates (n) operations for a list of size (n). Then, the deletion operation and the insertion operation are applied. This first phase requires (n+2) operations to identify the largest element, and then reproduce the same procedure on the new list of size ($n - 1$). Thus after K operations, this procedure is obtained:

$$K = (n+2)+(n+1)+(n)+(n-1)+\ldots+3 = \sum_{i=1}^{n} i+2 = \frac{n^2+5n}{2} \quad [2.1]$$

A second sorting method consists of using the notion of a "packet". A packet is associated with a quasicomplete well-ordered binary tree structure. A complete binary tree has $2^{h+1} - 1$ nodes, where h represents the level of the leaves (a leaf is a node that has no children and the level of a node is the distance). A quasicomplete binary tree may have any number of leaves. We explore the tree with a set of insertions and deletions of nodes. The tree is ordered if each node of the tree has a numerical value. We say that the tree is well ordered if the value of a parent is always less than that of its children (in the case of maximization or descending ordering). A packet is a structure that allows a new element to be added to a set of (n) elements in O(log(n)) operations, and allows the extraction of the minimum of the stack in O(log(n)) operations.

Finally, packet sorting a list of n elements begins with n insertions, followed by n extractions of the minimum value, for a total of $O(n \log(n))$ operations.

2.2.3. *Some definitions*

2.2.3.1. *Polynomial-time algorithms*

An algorithm is said to be of polynomial time if there exists a polynomial $Q(n)$ in n that is an upper bound of $T(n)$ for any numerical data:

$$T(n) \leq Q(n) \quad [2.2]$$

2.2.3.2. *Pseudo-polynomial-time algorithms*

An algorithm is said to be of pseudo-polynomial time if it is not of polynomial time, but there exists a polynomial which depends on the size of the problem n and another characteristic quantity P such that:

$$T(n) \leq Q(n, P) \quad [2.3]$$

2.2.3.3. *Exponential-time algorithms*

An algorithm is of exponential time if it is of neither polynomial nor pseudo-polynomial time.

2.2.4. *Complexity of a problem*

2.2.4.1. *Polynomial-time problems*

A problem is of polynomial time if there exists a polynomial-time algorithm to solve it exactly.

2.2.4.2. *NP-hard problems*

In an optimization problem, we seek the vector x^* in the set of solutions X such that the function $f(x)$ is minimized. We may define a decision problem associated with an optimization problem as, for example, does there exist a vector x in X such that $f(x) < A$? The answer to such a problem will be yes or no. The decision problems that cannot be solved in polynomial time are said to be NP-complete. The corresponding optimization problem is then considered to be NP-hard.

2.3. Exact methods

2.3.1. *Mathematical programming*

In mathematical programming, the objective of the studied problem is expressed in the form of a function of some variable (known as decision variables). This objective function is subject to a set of constraints of equality or inequality, which delimit the solution space to be explored. When the decision variables are integers and the functions are linear, this is an integer linear programming model. In the case of real integer variables, the model lies in the field of mixed integer programming. We also encounter problems of binary programming and nonlinear programming.

Different tools exist for obtaining the optimal solution to the problem from the mathematical model. Examples include the simplex method, dedicated solvers such as CPLEX®, XPRESS®, GAMS® and EXCEL®.

Various examples of mathematical models are given later in this book.

2.3.2. *Dynamic programming*

This method was introduced by Bellman in the 1950s [BEL 57]. Also called recursive optimization, it decomposes a given n-dimensional problem into n-unidimensional sub-problems. Thus, the system consists of n steps (or n decision levels) to be resolved sequentially. According to Chevalier, the transition from one step to another occurs according to evolution laws of the system and a decision [CHE 77]. This principle ensures that for every consecutive decision process, every sub-policy of an optimal policy is also optimal.

The use of dynamic programming to describe the optimal value of the criterion at a given step as a function of its value at the previous step depends on the existence of a recursive equation. From the initial combinatorial problem, the method consists of generating its sub-problems, solving them and determining the optimal trajectory. This is carried out by calculating a criterion for a subset k based on knowledge of this criterion for each subset $k - 1$, thus bringing the number of considered subsets to $2n$ (n being the number of elements considered in the problem). Owing to its exponential nature, which arises from the exponential generation of sub-problems, the method becomes very memory intensive.

Various examples of applying dynamic programming to the resolution of system design problems are given in Chapter 3.

2.3.3. *Branch and bound algorithm*

As discussed previously, it is not possible to produce a complete enumeration of every feasible solution for a given combinatorial optimization problem. There are, however, some techniques for the realization of an intelligent enumeration by exploring only certain subsets of solutions in which the optimal solution may be found. The basic idea is, therefore, to separate the set of solutions into subsets, and then evaluate them to see if the optimal solution may be contained therein. This approach, called a branch and bound (BAB) algorithm, requires the study of some properties (dominance properties and lower and upper bounds) of the problem so as to make use of tools to eliminate the bad solution subsets, which are said to be dominated.

DEFINITION 2.1.– *We define an upper bound as a value given by a heuristic (approximate solution). The upper bounds (UBs) are used as an initial solution in BAB algorithms. If we seek to minimize a criterion, then the smaller a given upper bound, the better it is.*

DEFINITION 2.2.– A lower bound (LB) is an optimistic estimate of the optimal solution S^* to the problem ($LB \leq S^*$ if we seek to minimize a criterion).

The application of a lower bound to a subset of solutions gives an idea of the quality of the solution found therein. If we seek to minimize a given criterion and we know one solution (an upper bound) S_1, then if the calculation of the lower bound of a solution subset gives a value of the criterion that cannot be less than $S_2 > S_1$; we may conclude that it is useless to explore this subset. The larger a given lower bound, the better it is (Figure 2.1).

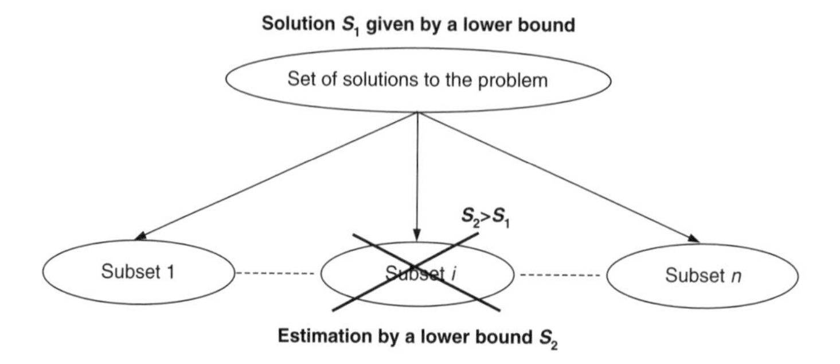

Figure 2.1. *The separation of the solutions into n subsets*

A dominance relation is a tool based on some properties of the problem, and allows comparison between the solutions (or families of solutions) S and S'. They are generally stated as follows: if we consider a solution S' constructed according to a property p', then the solution S, constructed according to a property p, will always be better. It is, therefore, useless to explore all the solutions S', which are said to be dominated by the solutions S.

2.4. Approximate methods

The optimal resolution of (NP-hard) optimization problems encountered in logistics and industrial management requires significant computation times, which are incompatible with the industrial requirements of operational management problems. Approximate methods are used to obtain acceptable solutions that are as close as possible to the optimal solution, but with an acceptable computation time. Unlike exact methods, which are intensive in both computation time and memory, and are limited as the size of the problem increases, approximate methods give importance to solution speed, especially in industrial environments where the time factor is important.

The methods to be presented are essentially based on metaheuristics known for their ability to solve NP-hard problems. These methods are simulated annealing [KIR 83], genetic algorithms [GOL 85], tabu search [GLO 90], ant colonies [DOR 92] and the particle swarm method. They constitute a family of optimization algorithms that are generally stochastic, and whose aim is to solve complex problems arising in operational research. They may be applied in different domains to solve various types of problems such as layout [HAN 06a], transport [DES 05], and vehicle routing [LAC 03].

2.4.1. *Genetic algorithms*

Genetic algorithms (GAs) are part of the family of evolutionary algorithms inspired by the biological evolution of species. They appeared toward the end of the 1950s [DRE 05]. They are based on Darwinian principles, in which the physical form of the individual determines its ability to survive and reproduce. The use of these algorithms to solve optimization problems was initiated by Goldberg [GOL 85, GOL 89].

2.4.1.1. *General principles*

An initial population is at first formed of n solutions (called chromosomes), selected randomly from the set of solutions. Each chromosome or individual i of the population is evaluated to measure its adaptation to the target objective, i.e. the quality of the solution thus encoded. Subsequently, the structure of the population changes by the application of progressive changes over several generations. During these K generations, we seek to improve the performance of the individuals, to make them evolve such that they are better adapted to survive. Thus, only the best-performing individuals are kept and selected to form the population of the next generation.

Individuals evolve due to different operators: crossover (the mixture of two parent chromosomes to obtain children), mutation (the random perturbation of a chromosome), and also sometimes local searches (we then speak of memetic algorithms).

2.4.1.2. *Encoding the solutions*

The first stage in the implementation of a genetic algorithm consists of encoding the solutions to the problem, i.e. constructing the encoding of the chromosomes so as to contain all the characteristic information about a solution. These elements constitute a list of genes represented by numbers. A set of genes then forms a chromosome that represents an individual or a possible solution.

Several encodings for different problems exist in the literature. Each problem has an appropriate encoding despite constraints on their construction. As an example, we present some studies involving the problem of scheduling in classic job-shop structures. In this type of problem, we must decide the order of a set of tasks in a workshop where each task uses a subset of the machine in a particular order:

– the Yamada *et al.* encoding [YAM 92] is based on a representation of the end dates of the operations for each task;

– the Tamaki encoding [TAM 92] uses a binary encoding translating the representation of the solution into a disjunctive graph;

– the Kobayashi *et al.* encoding [KOB 95] represents only the sequences of machine operations;

– the Portmann encoding [POR 96] proposes an indirect encoding in matrix form of the coordination of two consecutive operations;

– the Mesghouni encoding [MES 99] offers two variants that extend existing encodings. The first variant uses the Kobayashi encoding, by adding some information such as the job position and the starting time. The second, an extension of the Yamada encoding, integrates the machine at which the operations for each task are carried out; and

– the others such as the Kacem *et al.* encoding [KAC 01] and Nait-Tahar *et al.* encoding [TAH 04] may also be mentioned.

2.4.1.3. *Crossover operators*

The evolution of the population occurs through reproduction. Individuals are chosen to undergo crossover operations, which require mixing of two parent solutions to obtain one or two children. Crossover operators differ according to the chosen chromosome's encoding and the problem under consideration. It should be remembered that the crossover occurs with a certain given probability P_c, which may be static or varying during execution. Here, we give some examples of crossovers that may be encountered.

A 1X crossover operator (Figure 2.2) consists of randomly selecting a position in the parent chromosomes and creating two children by a succession of permutations.

Crossover operators are as numerous and varied as encodings. Next we present some examples of operators for a scheduling problem:

0	***1***	***1***	***0***	***1***	***1***	***0***	Parent 1
0	1	1	1	0	1	1	Parent 2
0	***1***	***1***	***0***	0	1	1	Child 1
0	1	1	1	***1***	***1***	***0***	Child 2

Figure 2.2. *A 1X crossover operator*

OX (Figure 2.3): It is a crossover with two crossover points and two children. Each of these four individuals is divided into three parts by randomly generating two crossover points along the length of the sub-chromosomes. The inner area of the mother is used for the inner part of the son, and the inner part of the father is used for the inner part of the daughter. Then, beginning with the third part and continuing with the first (from right to left on the chromosome), the missing values of the son are copied in the order of the father (for the daughter, the order of the mother is copied);

1	2	3	**4**	**5**	6	7	8	Parent 1
4	3	5	**6**	**7**	2	8	1	Parent 2
2	8	1	**4**	**5**	3	6	7	Child 1
4	5	8	**6**	**7**	1	2	3	Child 2

Figure 2.3. *A OX crossover operator*

LOX (Figure 2.4): It begins in a similar way to OX (regarding the cutting and filling of the crossover zone) but the filling of the two other zones is completed starting with the first part and ending with the third between the son, the father, the mother and the daughter (from left to right in the reading of its parents);

1	2	3	**4**	**5**	6	7	8	Parent 1
4	3	5	**6**	**7**	2	8	1	Parent 2
3	6	7	**4**	**5**	2	8	1	Child 1
1	2	3	**6**	**7**	4	5	8	Child 2

Figure 2.4. *A LOX crossover operator*

MT: It is a uniform crossover acting on the binary encoding of two children. We begin with two parents (mother and father), and we construct their respective precedence matrices: these are matrices composed of 0's and 1's, which are created by applying the following rule: if on a machine for scheduling (a sub-chromosome in our representation), task (i) precedes task (j), then the value of the precedence matrix for the pair (i,j) = 1, otherwise it is zero. To create a child (for example the son), we first determine its precedence matrix from that of the parents using the (boolean) logical operator "AND" between the matrix of the father and that of the mother. If, for a given pair, the value of the parent is different than that of the mother, then the value of the son will be equal to that of the father with a probability of (p) and equal to that of the mother with a probability of ($1 - p$) (and conversely in the case of the daughter). We then complete the matrix associated with the son by transitive closure (the precedence relation being transitive on the set of the groups). The different values of the father and mother not yet filled by transitive closure are successively evaluated and completed each time by transitive closure;

1X: It is a one-point crossover that, from two parents, gives rise to four children. It is in part a simplification of OX. The construction of the children is done as follows: for child 1 (child 2) we take the first part of parent 1 (parent 2) and complete the missing groups by taking them in the order of parent 2 (parent 1). For child 3 (child 4) we take the second part of parent 1 (parent 2) and finish the missing groups with their order in parent 2 (parent 1).

2.4.1.4. *Mutation operators*

Mutation consists of randomly taking genes in a child chromosome and modifying them. It is carried out with a relatively small probability and is therefore not systematically applied to every individual in the population. It generally gives a local optimum.

2.4.1.5. *Constructing the population in the next generation*

A selection for the replacement of the individuals of the population is applied. Different replacement procedures exist. The most common is replacing the bad individuals by children whose performances are better.

2.4.1.6. *Stopping condition*

The termination of the algorithm is directed by a criterion that may be a fixed number of generations, a computation time, a number of generations without improvement of the quality of the obtained results or a combination of these.

An illustration of the application of genetic algorithms to the resolution of a buffer sizing is given in Chapter 3.

2.4.2. *Ant colonies*

2.4.2.1. *General principle*

The principle of these methods is the mixture of *a priori* information on the structure of a promising solution and *a posteriori* information on the structure of recent good solutions. This type of algorithm consists of imitating the real behavior of ants in their parallel search for food across several paths based on the use of chemical substances called pheromones. The use of pheromones is considered as a dynamic memory structure that contains information about the quality of the different obtained solutions.

On the basis of this principle, the operation of ant colony algorithms may be summarized as follows. An ant colony, considered as a set of asynchronous computational agents, moves through different points that represent partial solutions to the problem. The movement of the ants is based on two stochastic decision parameters. These two parameters are the pheromone rate and the desirability rate. The desirability rate of a movement indicates the *a priori* tendency of an ant to make the movement. The pheromone level indicates the efficiency relating to the realization of a movement and thus constitutes *a posteriori* information about making this movement. As it moves, each ant will construct a solution to the problem. Once a solution is constructed, an evaluation of the solution and a modification of the pheromone levels are carried out. These pheromone updates will guide the future ants in their movements. A supplementary mechanism represented by the natural evaporation of the pheromones in time is applied so as to avoid an unlimited accumulation of pheromones in certain places. Only the good solutions may have increasing pheromone levels once all the ants have finished their rounds.

2.4.2.2. *Management of pheromones: example of the traveling salesman problem*

Ant colony algorithms were introduced by Dorigo [DOR 92] for the traveling salesman problem, in which we seek the shortest path that only passes once through the towns to be visited.

2.4.2.2.1. The construction of a solution

Let us consider a population of m ants and a graph consisting of n nodes to be visited. With each iteration t, $(1 \leq t \leq t_{max})$, each ant k $(k = 1, \ldots, m)$ constructs a complete path visiting n nodes. When an ant k is at node i, it may choose the following node j from the set J_i^k of nodes that it has not yet visited. The choice of node j is made by taking into account the probability associated with each node j:

$$p_{ij}^k = \begin{cases} \frac{(\tau_{ij}(t))^\alpha \times (\eta_{ij})^\beta}{\sum_{l \in J_i^k}((\tau_{il}(t))^\alpha \times (\eta_{il})^\beta)} & \text{if } j \in J_i^k \\ 0 & \text{otherwise} \end{cases} \quad [2.4]$$

This probability is calculated by taking into account heuristic information particular to the considered optimization problem. Here, this is the inverse of the distance between the nodes $\eta_{ij} = 1/d_{ij}$. This information is also called the visibility. The rate of pheromones $\tau_{ij}(t)$ also enters into this calculation. The parameters α and β modulate the significance of the intensity $\tau_{ij}(t)$ and the heuristic information η_{ij}.

2.4.2.2.2. Pheromone update

Once an ant has constructed a complete path, it leaves a certain quantity of pheromone $\Delta\tau_{ij}(t)$ between each pair of nodes i and j. This quantity depends on the quality of the solution obtained by the ant:

$$\Delta\tau_{ij}(t) = \sum_{k=1}^{m} \Delta\tau_{ij}^{k}(t) \quad [2.5]$$

$$\Delta\tau_{ij}^{k}(t) = \begin{cases} \frac{Q}{L^k(t)} & \text{if } (i,j) \in T^k(t) \\ 0 & \text{otherwise} \end{cases} \quad [2.6]$$

where $T^k(t)$ is the path constructed by ant k at iteration t, $L_k(t)$ is the length of the path, and Q is a fixed parameter representing the quantity of pheromones.

In order to avoid local optima, there is an evaporation principle given by the following formula:

$$\tau_{ij}(t+1) = \rho\tau_{ij}(t) + \Delta\tau_{ij}(t) \quad [2.7]$$

where ρ is the rate of evaporation $(0 < \rho < 1)$.

Algorithm 2.1 summarizes the implementation of ant colony optimization for the traveling salesman problem.

Algorithm 2.1 Ant colony optimization algorithm

1: $t = 1, ...t_{max}$
2: **for** each ant $k = 1, ...m$ **do**
3: randomly choose a node
4: **while** $J_i^k \neq \emptyset$ **do**
5: choose a node j in J_i^k with the corresponding probabilities
6: J_i^k = J_i^k - $\{j\}$
7: **end while**
8: update the pheromones with the corresponding equations
9: **end for**

An example application of ant colony algorithms is presented in Chapter 3 to resolve a buffer sizing problem.

2.4.3. *Tabu search*

Tabu search is an combinatorial search algorithm that uses a tabu (taboo) list to perform an intelligent search. This algorithm consists of a guided search from an initial solution found with a relatively simple and effective method. The search is performed following movements around a solution, and constructs a set of new neighboring solutions. A neighboring solution is a solution that was constructed from another solution, and that consequently retains a certain number of the characteristics of its progenitor.

A generational loop lets us avoid local optima. In the following section, we present the main elements that make up the algorithm, as well as an example of its usage in the resolution of a scheduling problem.

2.4.3.1. *Initial solution*

A tabu search requires an initial solution. This solution is found using simple methods such as list algorithms. Given that the performance of a tabu search can be strongly dependent on the performance of this solution, it is important that the initial solution has an acceptable performance. The time required to construct the initial solution is also an important factor to consider so as not to increase the execution time of the tabu method.

In the solution of scheduling problems, the initial solutions are often constructed with list algorithms that, according to well-known rules such as the shortest processing time (SPT) or others, assign priorities to tasks and allocate them on the set of available machines.

Then, the initial solution enters into the generational loop. For every solution to be correctly identifiable, it is necessary to set a representation of the solution common to every solution.

2.4.3.2. *Representing the solution*

To perform a generational loop-type iterative search, we must define a uniform representation of the solution. This representation must provide all the information necessary to reconstruct and evaluate each solution. A list or vector can normally represent the solutions. In certain cases, several lists or tables are used. If special constraints are taken into account, other rules may be added to complete the description of the solution.

The most often-used form of representation is the priority list. A priority list is a vector that gives the priority of each element in the list with respect to its position. For example in scheduling problems, tasks are allocated to machines depending on their position in the vector that represents the solution. An example of a solution representation in a scheduling problem is given in Figure 2.5.

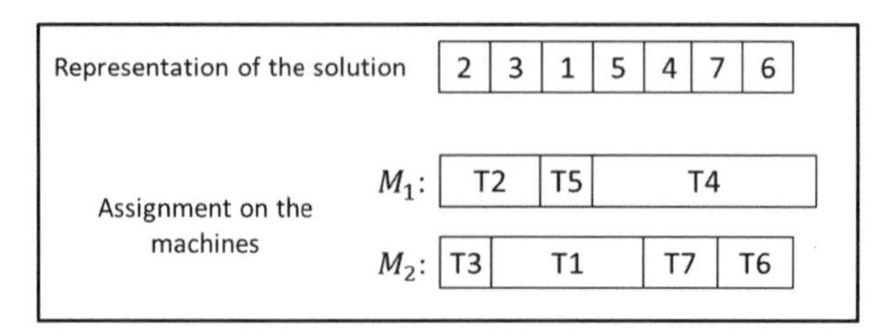

Figure 2.5. *Representing the solution: an example of scheduling seven tasks on two machines*

Each solution must have a unique representation, and each representation must correspond to a single solution. Any ambiguity may lead to infinite loops.

2.4.3.3. *Creating the neighborhood*

From a given solution, which we will call the seed solution, new solutions are constructed by modifying its composition. Movements around the representation of the seed solution are carried out to find different solutions. These solutions may be new, or may have already been seen. To avoid returning to previously constructed solutions, a tabu list is used.

The set of solutions generated from the seed solution is the neighborhood of this solution. This neighborhood is then tested against the chosen optimization criterion. In certain cases, measuring this criterion on every neighbor of a solution proves to be a very complicated task. A candidate list is then used. This list is a simplified procedure that weighs the potential of a solution with respect to a more easily measurable criterion. Such criteria are relaxations of the original problem, or approximations of a solution's performance compared with the seed solution.

As we have already specified, to construct a neighborhood, we carry out movements around the seed solution. These movements are iterations on its representation. Among the most common movements, pairwise exchange is regularly used due to its ease of application. The example in Figure 2.6 shows the pairwise exchange of two elements of a solution. Pairwise exchange is an intensification technique, which means that the solutions found will be very close to the seed solution.

Other types of movement are used to escape from sub-optimal solutions: diversification movements. One example of this type of movement is mutation. When we carry out a mutation movement in the seed solution, its composition is significantly altered. The aim is to introduce very different elements, and to explore new possibilities. An example of mutation is shown in Figure 2.7.

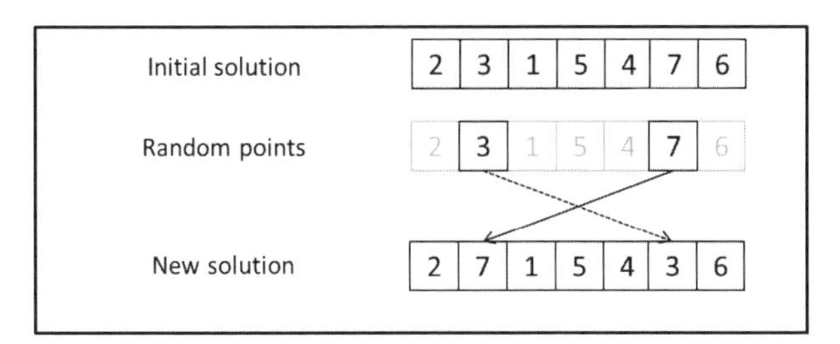

Figure 2.6. *Pairwise exchange: an example with seven tasks and two random points*

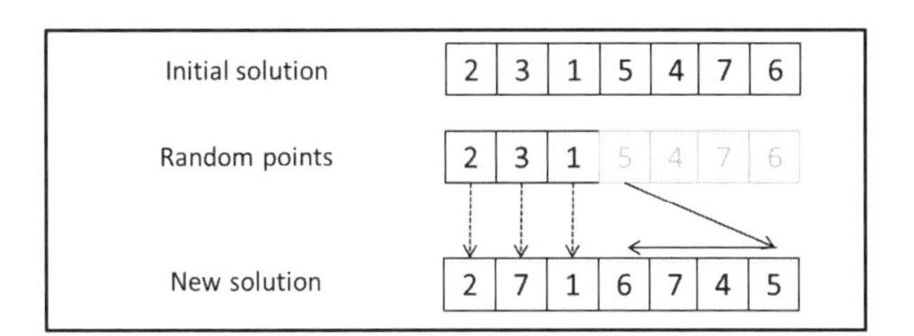

Figure 2.7. *Mutation: an example with seven tasks and one random point*

2.4.3.4. *The tabu list*

The tabu list is at the core of this method. This list is a table that prevents the occurrence of movements which will probably lead to previously visited solutions (a visited solution is the solution that has already been part of a created neighborhood). This list comes into action during the creation of a neighborhood. It compares the movements that will be carried out with the ones in the tabu list. If a movement is considered tabu (i.e. present in the tabu list), it cannot be executed as long as it is not deleted from the list.

It is necessary to define an inclusion policy in the tabu list. In scheduling problems, it is common to find this type of inclusion policy: when a movement is carried out and produces the best solution of the neighborhood (which is taken as the seed solution of the next neighborhood), the movement of the elements that participated in it becomes tabu. If we integrate several types of movement, the tabu list must allow the identification of not only the included elements, but also the type of movement used.

To complete the description of the tabu list, it is imperative that the duration of the period a movement remains tabu is defined. This value is often a random variable that is distributed across a fixed interval.

Algorithm 2.2 shows the structure of a tabu search.

Algorithm 2.2 A tabu search algorithm

1: $S_0 \leftarrow Initial\ solution$
2: $Best \leftarrow f(S_0) \leftarrow Performance\ of\ the\ initial\ solution$
3: **for** $k = 1$ to the number of generations **do**
4: **for** $j = 1$ to $Size\ of\ neighborhood$ **do**
5: $S_j \leftarrow New\ solution\ from\ S_0$
6: $f(S_j) \leftarrow Performance\ of\ the\ new\ solution$
7: **if** $f(S_j) < LocalBest$ and $S_j \notin TabuList$ **then**
8: $LocalBest = f(S_j)$
9: $S_{Best} \leftarrow S_j$
10: **end if**
11: **end for**
12: $S_0 \leftarrow S_{Best}$
13: $TabuList \leftarrow Moves(S_{Best})$
14: **end for**
15: $Best\ solution = S_{Best}$

2.4.3.5. *An illustrative example*

The example that we present is a classic scheduling problem. The problem consists of finding an order for a set of tasks on a group of identical machines. Each task has a processing time and a desired end date. The objective is to minimize the total tardiness (delay), i.e. the sum of the delay of all the tasks once they have been allocated to the machines. The delay is the time that elapses between the task's desired end date and its real end date.

The solution is represented as a list of task priorities. These priorities will be carried out according to the order they appear in the list. We add the allocation rule which says that a task will be treated by the first available machine.

The initial solution is constructed according to the SPT method, where the priority of the task is given by its processing time. Thus, the task that has the SPT has priority during the allocation of tasks to machines.

2.4.4. *Particle swarm algorithm*

2.4.4.1. *Description*

Particle swarm optimization (PSO) is a stochastic optimization technique based on population search algorithms. It was first used by Eberhart and Kennedy in 1995 [KEN 95]. This method reproduces the behavior of swarms of bees or flocks of birds.

The founding principle is the specific management of information exchanged between the particles to decide their next move.

This approach is based on swarm intelligence in which each particle represents a solution to the problem being studied. A particle moves in the solution space according to its velocity, its memory, and some informants. The algorithm starts by randomly initializing the positions and velocity of the particles. After some iterations, the future particles aim to find a good solution using the best solution found by their autonomous movements (P_{best}) and the best global solution found by the particle (G_{best}). This is weighted by a random number. The difference between (P_{best}) and (G_{best}) is that the latter is used for every particle and the former is used for an individual particle.

The movement of the particles is based on three pieces of information: velocity, the best position and the best position of the particles, of the neighboring particles. This information is weighted using $\theta 1$, $\theta 2$, and $\theta 3$. This allows the significance of these three parameters to be modulated, and it is therefore important to choose the correct values. This choice prevents the particles from being attracted by the best position that may lead us to fall into a local optimum.

The particle's choice of trajectory is based on three pieces of information: the trajectory of the swarm leader, its best trajectory, and its velocity.

The position of the leader represents the best value of the function obtained from the start of the swarm (the start of the algorithm) up to its current position (generation t) found among all of the particles. The best trajectory of the particle represents the best solution obtained by the particle itself from the start to its current position. The particle may also follow its own trajectory, based on the inertia resulting from its velocity.

2.4.4.2. *An illustrative example*

The method is of recent success and has increasingly wide applications. Yalaoui [YAL 10] applied PSO to solve *flow shop* scheduling problems. Each solution contains the positions of the tasks (the sequence), as described in Table 2.2. For example at stage 1, task 2 is in first position followed by task 1 and finally task 3. The same logic is applied to the two other stages. This sequencing is obtained mainly using the position X_p^t of the particle. In this case, this allows the sorting of tasks according to their increasing values at each iteration. In the example, at stage 1, task 2 is found in first position due to its position having the smallest value, and so on for the rest of the tasks.

The standard version of the algorithm gives a maximum velocity that is equal to the size of the particle. In the example, we must order three tasks so that V_{max}

(the maximum velocity that is equal to the number of tasks) equals 3, and the same principle is applied to the positions. The value of the velocity represents the movements that occur for each task. This is done in the following way:

$$V_p^{t+1} = (\theta 1.V_p^k + \theta 2.rnd(p).(X(P_{best}) - X_i^k) + \theta 3.rnd(g).(X(G_{best}) - X_i^k))modV_{max} \quad [2.8]$$

$$X_p^{t+1} = X_p^k + V_p^{t+1} modV_{max} \quad [2.9]$$

Notation:

– $\theta 1$: the inertia;

– $\theta 2$: the parameter giving the significance of P_{best};

– $\theta 3$: the parameter giving the significance of G_{best};

– rnd(p): a random number in the interval [0,1];

– rnd(g): a random number in the interval [0,1];

– X_p^{t+1}: the future position of the updated particle;

– X_p^k: the current position of the particle;

– V_p^{t+1}: the velocity of the updated particles; and

– V_{max}: the maximum velocity, which is equal to the number of tasks.

	Stage 1			Stage 2			Stage 3		
Sequence	2	1	3	2	1	3	1	2	3
X_p^t	0.223	0.329	0.758	0.234	0.574	0.864	0.453	0.456	0.987
V_p^t	0.223	0.329	0.758	0.234	0.574	0.864	0.453	0.456	0.987

Table 2.2. *An example of particle encoding in PSO*

In this example, we use another method to update the positions, and we take as an example the solution from Table 2.2 at iteration t. The values taken by the velocity vary between 0 and the maximum number of tasks to be treated. The new values of the velocity for iteration $t + 1$ are calculated according to equation [2.8]. The results are presented in Table 2.3.

The position updates are made according to equation [2.9] (Table 2.4). The sequence updates are made by sorting new positions in ascending order, which allows a new task order to be constructed (Table 2.5).

	Stage 1			Stage 2			Stage 3		
Sequence	2	1	3	2	1	3	1	2	3
X_p^t	0.223	0.329	0.758	0.234	0.574	0.864	0.453	0.456	0.987
V_p^t	2.578	1.298	0.222	2.120	1.994	0.729	0.027	0.876	0.490

Table 2.3. *Velocity updates in PSO*

	Stage 1			Stage 2			Stage 3		
Sequence	2	1	3	2	1	3	1	2	3
X_p^t	2.801	1.627	0.970	2.354	2.468	1.585	0.480	1.322	1.477
V_p^t	2.578	1.298	0.222	2.120	1.994	0.729	0.027	0.876	0.490

Table 2.4. *Position updates in PSO*

	Stage 1			Stage 2			Stage 3		
Sequence	3	1	2	3	2	1	1	2	3
X_p^t	0.970	1.627	2.801	1.585	2.354	2.468	0.480	1.322	1.477
V_p^t	0.222	1.298	2.578	0.729	2.120	1.994	0.027	0.876	0.490

Table 2.5. *Sequence updates in PSO*

2.5. Multi-objective optimization

2.5.1. *Definition*

Multi-objective optimization occupies an important place in current fields of research into real engineering problems in the industrial world. It allows several, often contradictory, criteria to be simultaneously optimized. Thus, we no longer seek to determine an optimal solution, but instead seek a set of compromise solutions. A multi-objective optimization problem consists of finding a vector X^* (where $X^* = [x_1^*, x_2^*, \ldots, x_n^*]^T$) of decision variables that satisfy the m inequalities [2.10] and the p equations [2.11] in order to optimize the vector $F(X)$ of function [2.12]:

$$g_i(x) \geq 0, \; i = 1, 2, \ldots, m \qquad [2.10]$$

$$h_i(x) = 0, \; i = 1, 2, \ldots, p \qquad [2.11]$$

$$F(X) = [f_1(X), f_2(X), \ldots, f_k(X)]^T \qquad [2.12]$$

In other words, a particular set $x_1^*, x_2^*, \ldots, x_n^*$ giving a compromise for all of the objective functions that satisfy the constraints [2.10] and [2.11] must be found. This compromise is represented by a set of non-dominated solutions grouped into an optimal front. The dominance of a solution over another is determined according to certain techniques that we will develop later. None of the solutions in the optimal front may be considered as the best solution to the problem. The choice of which solution to apply requires a precise understanding of the studied problem and a given number of additional factors [SRI 94].

2.5.2. *Resolution methods*

Different methods are used in the literature for solving multi-objective optimization problems [MAK 08b].

One of these approaches consists of aggregating all of the objectives, thus creating a single-criterion optimization problem. For this, we calculate a linear weighted sum [ISH 03, HAN 06c]. However, modeling the problem in the form of a single equation may be a difficult task. Also, seeking to attain a single-objective function may bias the model [COL 02].

Another solution method consists of creating different populations in genetic algorithms. A single objective is associated with each population [FON 95].

In 2007, another approach based on Lorenz dominance has proved its efficiency in certain applications [DUG 07]. Lorenz dominance requires a more significant computation time for each solution, but on the other hand it reduces the size of the optimal front. This is why the Lorenz optimal front is a subset of the Pareto optimal front [DUG 08].

Nonetheless, the approach most often used to solve multi-objective optimization problems continues to be that of Pareto. We present this approach in the following section.

Pareto dominance [PAR 96] is one of the main methods for solving multi-objective problems. The Pareto dominance relation considers the optimal solutions of a multi-objective problem to be those that are non-dominated and grouped into a single front.

For a maximization problem, we say that a solution A is non-dominated, or dominates another solution B in the Pareto sense, if and only if, for any objective i (i belongs to the set of k considered objectives), the solution A gives either the same or a better performance than the solution B (equation [2.13]), and for at least one objective j, the solution must be the best solution (equation [2.14]):

$$\forall i \in \{1, 2, \ldots, k\} : \; f_i(A) \geq f_i(B) \qquad [2.13]$$

$$\exists j \in \{1, 2, \ldots, k\} : \; f_i(A) > f_i(B) \qquad [2.14]$$

The set of solutions that dominate the others but do not dominate each other are called Pareto optimal, Pareto efficient, or non-dominated solutions, and they constitute the Pareto front.

2.5.3. *Comparison criteria*

To compare fronts of non-dominated solutions obtained by multi-objective optimization algorithms, we can use several metrics. This comparison is important since it will allow the quality of obtained solutions to be evaluated, and allow us to determine which algorithm is more advantageous than another. These criteria are divided into two categories:

– criteria specific to one front:

- the number of solutions in the optimal front n_f,

- the hypersurface and the hypersurface ratio [COL 02],

- the spacing [COL 02], and

- the HRS metric [COL 02].

– criteria specific to two fronts:

- the distance μ proposed by Riise [RII 02],

- the Zitzler measure [ZIT 99],

- the progression measure [COL 02], and

- the Laumanns metric [LAU 00].

In the following sections, we will describe the Riise distance and the Zitzler measure. These two criteria have been efficiently applied in several works [AMO 07, DUG 08, LAC 06].

2.5.3.1. *The Riise distance*

The Riise distance μ [RII 02] is calculated according to equation [2.15]:

$$\mu = \sum_{x=1}^{nf} d_x \qquad [2.15]$$

Let F_1 and F_2 be two fronts obtained with two multi-objective optimization algorithms in the case of a problem where the two objectives are to be minimized.

In equation [2.15], d_x is the distance between a solution x belonging to front F_1 and its orthogonal projection onto front F_2. The distance μ is negative if F_1 is below F_2, and positive otherwise. Given that μ depends on the number of solutions n_f in front F_1, there is generally a normalization according to equation [2.16]:

$$\mu^* = \frac{\mu}{n_f} \qquad [2.16]$$

2.5.3.2. *The Zitzler measure*

Let F_1 and F_2 be two fronts to be compared. The Zitzler measure [ZIT 99] (C_1) represents the percentage of solutions in F_1 that are dominated by at least one solution in front F_2. Given that this measure is not symmetric, it is necessary to also calculate (C_2). The latter therefore corresponds to the percentage of solutions in F_2 that are dominated by at least one solution in front F_1. In conclusion, F_1 is of better quality than F_2 if $C_1 \leq C_2$.

2.5.4. *Multi-objective optimization methods*

In the following sections, we present, in detail, the main exact and approximate methods for solving multi-objective optimization problems.

2.5.4.1. *Exact methods*

This section presents the most widely used exact methods of solving multi-objective optimization problems.

First of all, it must be noted that a large number of methods essentially combine multiple objectives into a single method. The most popular is the aggregation method (or the weighted sum). This method transforms a multi-objective problem into a single-objective problem based on the following criterion:

$$Optimize \sum_{i=1}^{n} \lambda_i f_i(x) \qquad [2.17]$$

where $0 < \lambda_i < 1$ and $\sum_{i=1}^{n} \lambda_i = 1$. The supported solutions may be found by varying the weighting factors. The advantage of this method (in problems where the single-objective version may be resolved in polynomial time) is that for each λ, the problem is always a single-objective problem; the difficulty of the problem is reduced. However, this method is not capable of finding non-supported solutions.

2.5.4.1.1. The ε-constraint method

In order to list all of the Pareto optimal solutions, the ε-constraint principle [HAI 71] is used in this method. The multi-objective optimization problem is thus transformed into a single-objective optimization problem. A single objective is considered while others are accounted for as constraints. This implies that each time an optimal solution is found (with a single objective), an additional constraint is added for the other objective. In other words, the search always begins by finding the extreme solution that has the best value for the first objective $f_1(x)$. When a solution U is found, the next search is to find another solution x that has the best value of $f_1(x)$ with the constraint $f_2(x) < f_2(U)$. This process is repeated until new solutions are found. In this way, we obtain all of the Pareto optimal solutions or the absolutely optimal Pareto front.

Applications of this method may be found in the works of Laumanns *et al.* [LAU 06] and Mavrotas [MAV 09], for example.

2.5.4.1.2. The two-phase method

Proposed by Ulungu and Teghem [ULU 95] in 1995, this method is used to find all of the Pareto optimal solutions. The Pareto optimal solutions are classed into two categories: supported solutions and non-supported solutions. The identification of the supported solutions is done based on the aggregation method, using different combinations of λ to find all of the solutions situated in the convex set of the Pareto front. The second step consists of finding the non-supported solutions in the triangular zones composed of the adjacent supported solutions.

As examples, we cite the work of Lemesre *et al.* [LEM 07a], solving a flow shop scheduling problem, and that of Przybylski [PRZ 06], solving a combinatorial multi-objective optimization problem with the two-phase method.

2.5.4.1.3. The parallel partitioning method

Proposed in 2007 by Lemesre *et al.* [LEM 07b] to solve a two-criteria optimization problem, the parallel partitioning method (PPM) consisting of three main steps is used. The first step is the search for extreme solutions. The second step is dividing the search space into different zones. All of the remaining Pareto optimal solutions are then found in the third step. In order to reduce the search space, a low computation time is required for the application of this method. As an example, we may cite the work of Lemesre *et al.* [LEM 07b] who have applied this method in a flow shop scheduling problem.

2.5.4.1.4. The K-PPM method

Until now, the exact multi-objective methods presented have been applied to two-objective optimization problems. A new method designed to solve problems with k

objectives is proposed by Dhaenens *et al.* [DHA 10]. Called the K-PPM method, and based on the PPM method, it consists of dividing the search space into several domains. There are three steps. The first step in this method consists of finding the best and worst points with regard to the considered objectives (*Ideal and Nadir points*). Once these points are found, the set Opt^{k-1} is identified (similar to seeking extreme points in the two-objective case). The second step consists of searching for certain well-distributed solutions to divide the search space. Finally, the third step consists of finding all the other solutions while exploring the search subsets. This method proposes an interesting division of the search space by finding distributed solutions.

2.5.4.2. *Approximate methods*

Exact methods are generally limited in their capacity to solve large, complex problems in polynomial time. Approximate methods are therefore increasingly used to solve multi-objective optimization problems. These methods may be divided into two large families: heuristics and metaheuristics.

Heuristic is defined as being a technique that seeks good solutions at reasonable cost without being able to guarantee optimality [REE 93]. This type of method is often used to solve problems with a particular nature, and is not always well-adapted to solve another problem.

Metaheuristic, on the other hand, is a powerful techniques that may be applied to a large number of problems of different types. It refers to a master strategy that guides and modifies the operations of subordinate heuristics by intelligently combining different concepts to explore the search space [OSM 96]. It is based on the ability to manipulate a single solution or a collection of solutions at each iteration. Therefore, because of the flexibility of metaheuristics, much research has been conducted on the application of metaheuristics to multi-objective optimization problems [COE 07].

Genetic algorithms show an interesting performance concerning the resolution of multi-objective optimization problems [DAN 06]. Recently, we have seen the development of the first algorithms based on ant colony optimization but applied to multi-objective optimization problems, such as in the work of Benlian and Zhiquan [BEN 07] and Pellegrini *et al.* [PEL 07].

2.5.4.2.1. Multi-objective genetic algorithms

Several types of methods based on genetic algorithms may be used to solve multi-objective optimization problems. These methods are generally based on the notion of Pareto dominance. In the following sections, we present some multi-objective metaheuristics based on genetic algorithms.

2.5.4.2.2. NSGA

This multi-objective algorithm, known as a non-dominated sorting genetic algorithm, was proposed by Deb and Srinivas [DEB 94]. Its structure is as follows:

– From the initial population, the non-dominated solutions or individuals are sought according to the notion of Pareto dominance. The set of these solutions constitutes the first Pareto front.

– A *fitness* value is attributed to each solution. This value gives every solution an equal chance of reproduction. Furthermore, to guarantee diversity in the population, a sharing function is applied.

– The first group of individuals is then deleted from the population.

– The three steps above are repeated to determine the second Pareto front. The fitness value attributed to this second group is less than the smallest fitness value of the first group after the sharing function was applied to it. This mechanism is repeated until all of the individuals in the population have been considered.

The next step in this algorithm is identical to a classic genetic algorithm (selection, crossover, mutation, etc.). To select individuals, Deb and Srivinas [DEB 94] used a selection method based on the stochastic remainder, although other selection heuristics may be used (tournaments, roulette, etc.). The advantage of this method lies in its use of a sharing function, which gives a diverse population and solutions that are more efficiently distributed in the Pareto fronts. This method may be equally applied to any multi-objective problem, regardless of the number of criteria to optimize.

2.5.4.2.3. NSGA-II

Deb *et al.* [DEB 02] proposed the second version of the NSGA algorithm: NSGA-II. In this version, the authors attempted to improve the efficiency of the NSGA algorithm by removing certain aspects such as its complexity, its non-elitism, and the use of the sharing function.

The complexity of the NSGA algorithm is notably due to the process of creating the different fronts. To overcome this problem, Deb *et al.* proposed a modification of the procedure for sorting the population into several fronts. To do this, they used a neighborhood density for the solutions. Finding the density of a solution i consists of calculating the mean distance between the considered solution and the two closest points on either side of the solution. This distance is calculated for each studied objective. It is called the crowding distance, and will be used to guide the process of selecting individuals.

To address the second criticism of the first version of the algorithm, i.e. not using elitist selection, the authors used selection by tournament, and modified the procedure for moving between generations. If two solutions are chosen to participate in the tournament, the solution of lowest rank is retained (the rank is related to the number of the front where the solution is situated). If, on the other hand, the two ranks are identical, it is better to use the farther solution, i.e. the rank with the higher crowding distance.

Compared to the first version, the NSGA-II algorithm has the following advantages:

– an elitist approach that stores the best solutions;

– a faster sorting procedure; and

– a comparison operator based on the calculation of the crowding distance.

2.5.4.2.4. The Strength Pareto Evolutionary algorithm

The Strength Pareto Evolutionary algorithm (SPEA) is a multi-objective evolutionary algorithm proposed by Zitzler and Thiele [ZIT 98] in 1998. Based on the notion of Pareto dominance, this algorithm uses an external population called an *archive*, which consists of recording the elite solutions. The archive may therefore contain a limited number of non-dominated solutions during the different generations of the algorithm. At each iteration, new non-dominated solutions are compared with the elite solutions in the archive. Only the resulting non-dominated solutions are kept in the archive. To preserve diversity, we can use a new nesting technique based on Pareto dominance.

The transition from one generation to another begins by updating the archive. All the non-dominated individuals are copied into the archive and the dominated individuals are deleted. If the number of individuals in the archive exceeds a given number, the clustering technique is applied to reduce the archive. The fitness value of each individual is updated before carrying out the selection using the two sets of solutions from the two generations. Finally, modification genetic operators are applied.

This method efficiently distributes the solutions on the Pareto front. The scoring technique leads to a good sample of the individuals in the space. The concept of strength associated with the clustering technique leads to the creation of nests in which the score of the individuals depends on their position relative to the Pareto optimal individuals.

2.5.4.2.5. SPEA-II

Zitzler and Thiele [ZIT 01] proposed a second version of the SPEA algorithm. As in the first version, this algorithm is based on the use of a population and an archiving mechanism. At the end of each generation, all the non-dominated solutions are copied into the archive. The difference from the SPEA algorithm is that the size of the archive is fixed in this algorithm. The archive is therefore filled by non-dominated solutions up to the fixed size. In the case where the number of non-dominated solutions is less than the size of the archive, the dominated solutions that have the best objective functions are selected. If the number of non-dominated solutions is greater than the size of the archive, the best solutions are kept in descending order relative to the distance σ, which calculates the distance separating each solution from the others of the population. Compared to the first version, the SPEA-II algorithm has the following advantages:

– a new, more efficient fitness function is adopted;

– the density of solutions is incorporated into the calculation of performance; and

– a new archive is applied to guarantee the preservation of the frontier solutions.

2.5.4.2.6. Multi-objective ant colonies

The principle of ant colony systems is based on the deposition of chemical substances (pheromones) by ants on their paths toward a source of food. These optimization algorithms are therefore based on pheromones, using matrices of quantities deposited between solutions that thus allow the optimal solution to be found. Given that multi-objective optimization takes several criteria into consideration, we consider the total number of pheromone matrices to be equal to the number of objective to be optimized.

2.5.4.2.7. Constructing the ant trails

Once the encoding of the solutions has been constructed (according to the type of problem), the construction of the ant trails is carried out in the following way. First, each ant is randomly placed on an initial point corresponding to the size of the first buffer. Then, each ant constructs its path by choosing the next points to visit.

Each point on the path therefore represents a sub-solution to the problem. The choice of points to visit is based on the application of the state transition rule. An ant k at a point r chooses the next point s to visit according to equation [2.18]:

$$s = \begin{cases} argmax_{u \in J_k(r)} \left\{ \left[\sum_{o=1}^{O} w_o . \tau_{r,u}^{o} \right]^{\alpha} . \left[\eta_{r,u} \right]^{\beta} \right\} & \text{if } q \leq q_0 \\ S^* & \text{otherwise} \end{cases} \quad [2.18]$$

In equation [2.18], q is a randomly generated number between 0 and 1, q_0 is a parameter ($0 \leq q_0 \leq 1$) that determines the relative importance of exploitation compared to exploration. S^* is a random variable chosen according to a probability given by equation [2.19]. O is the number of objectives considered.

Elements o, $\tau^o_{r,s}$, and w_o represent, respectively, the quantity of pheromones between points r and s and the weighting factor assigned to each objective. $\eta_{r,s}$ is a static value used as heuristic of the innate desirability or the visibility of choosing s from r.

The calculation of the visibility depends on each studied problem, and may be considered as a heuristic value. The parameters α and β are used to give more significance to either the visibility or the quantity of pheromones. $J_k(r)$ is the set of points not yet visited by ant k:

$$S^* = \begin{cases} \dfrac{\left[\sum_{o=1}^{O} w_o.\tau^o_{r,s}\right]^\alpha . [\eta_{r,s}]^\beta}{\sum_{u \in J_k(r)} \left[\sum_{o=1}^{O} w_o.\tau^o_{r,u}\right]^\alpha . [\eta_{r,u}]^\beta} & \text{if } s \in J_k(r) \\ 0 & \text{otherwise} \end{cases} \quad [2.19]$$

2.5.4.2.8. Global and local updates of pheromones

In constructing its path, an ant changes the quantity of pheromones on the visited arcs by applying a local update function according to equation [2.20], where ρ is the rate of pheromone evaporation ($0 \leq \rho \leq 1$) and τ_0 is the initial amount of pheromones:

$$\tau^o_{r,s} = (1 - \rho).\tau^o_{r,s} + \rho.\tau^o_0 \quad [2.20]$$

τ_0 is calculated according to the particular problem. Let us take as an example the problem of buffer sizing, with the aim of minimizing the total size of the buffers and maximizing the performance rate of the line. Equation [2.21] shows the initial quantity of pheromones for the pheromone matrix of the first objective (minimization of the total size) and equation [2.22] for the second objective (maximization of the output rate). We then choose to initialize the pheromones according to the maximum capacities of the buffers:

$$\tau^1_0 = \left(U_{BC}.\sum_{i=1}^{N-1} b_{ui}\right)^{-1} \quad [2.21]$$

$$\tau^2_0 = N_{ps\ max} \quad [2.22]$$

Once all the ants have finished their turn, the quantity of pheromones on each arc is modified according to a global update function, as shown in equation [2.23]:

$$\tau^o_{r,s} = (1 - \rho).\tau^o_{r,s} + \rho.\Delta\tau^o_{r,s} \quad [2.23]$$

$\Delta\tau^o_{r,s}$ is a factor that favors the non-dominated solutions. It is calculated according to equations [2.24] and [2.25]:

$$\Delta\tau^1_{r,s} = \begin{cases} (C_{gb})^{-1} & \text{if arc } r, s \text{ belongs to a non-dominated solution} \\ 0 & \text{otherwise} \end{cases} \quad [2.24]$$

$$\Delta\tau^2_{r,s} = \begin{cases} P_{gb} & \text{if arc } r, s \text{ belongs to a non-dominated solution} \\ 0 & \text{otherwise} \end{cases} \quad [2.25]$$

C_{gb} and P_{gb} represent the minimum total size and the maximum performance rate, respectively obtained so far by the ants. With these updates to the pheromone matrices, the arcs with the highest quantities of pheromones will attract more ants in the next step.

2.6. Simulation-based optimization

Simulation-based optimization may be defined as the pairing of an optimization method with a simulation model to test several parameters that may maximize the performance of the system being studied. This technique may be carried out by coupling the simulation module with a dedicated optimization module, or programming a new module based on the particular case under consideration.

Figure 2.8 shows the concept of this technique. The optimization module communicates the values of the decision parameters to the simulation module. These parameters may be queue management policies of the machines in a system, the buffer capacities in a line, the locations of machines in a workshop, etc. The simulation module allows the evaluation of the solutions with the proposed decision parameters.

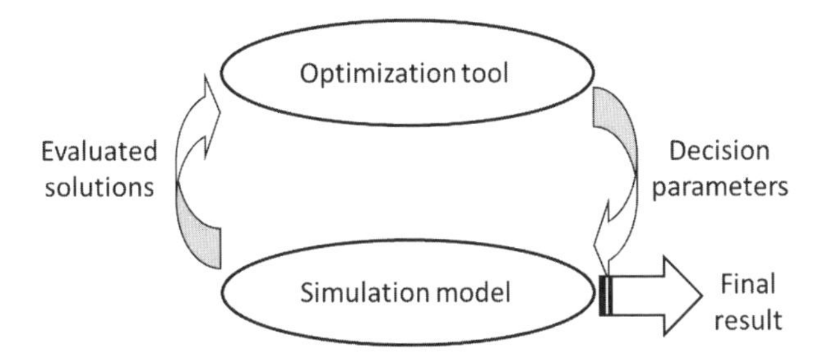

Figure 2.8. *Simulation-based optimization*

The results are then sent to the optimization tool to carry out the different steps of the adopted algorithm (selection, crossing and mutation for genetic algorithms, global and local updates for ant colonies, etc.) and propose new decision parameters to be evaluated. These steps are repeated until a termination criterion is satisfied. Once this is the case, we may obtain the final result, which may be the optimal solution or the best solution found to the problem being studied.

Simulation-based optimization has been the subject of much work. In 1998, Elmaraghy *et al.* [ELM 98] applied this technique for minimizing the *makespan* in a flexible production system. In 2003, Muhl *et al.* [YAN 03] used simulation coupled with genetic algorithms and simulated annealing to solve a planning problem in the automotive industry.

2.6.1. *Dedicated tools*

Several dedicated tools and softwares, such as Autostat, have appeared on the market in recent years. They may be directly coupled with simulation programs as Glover *et al.* [GLO 96] showed in 1996.

Among the tools that may be coupled with the ARENA® software, we find OptQuest®, which involves an optimization algorithm based on tabu search. Several works have shown the objectives and the effects of pairing ARENA® with OptQuest®, such as those of Sadowski and Bapat [SAD 99], Fu [FU 02], and Yang *et al.* [YAN 05].

CPLEX® is another tool developed to solve mathematical programming problems. The logistics system studied by Vamanan *et al.* [VAM 04] in 2004 presents a method of pairing ARENA® with CPLEX®.

2.6.2. *Specific methods*

The second way of applying simulation-based optimization is the use of specific methods based, in general, on heuristics or metaheuristics such as genetic algorithms [LIN 97] and ant colonies [GAG 02].

The coupling of these methods with simulation software such as ARENA® has been the subject of several works, such as those of Harmonosky [HAR 95] and Drake and Smith [DRA 96]. This coupling optimizes the models simulating different systems, and may solve single-criterion or multi-criteria optimization problems.

This coupling allows the optimization of the simulation of different systems to solve various optimization problems, such as shown in the work of Liu and Wu [LIU 04] to determine the best scheduling policy, or in that of Zhang and Li [ZHA 04] for dynamic resource allocation. In the work of Hani *et al.* [HAN 06c], a genetic algorithm is coupled with a simulation model in ARENA® for the optimal choice of queue management policy for railway maintenance sites. In another work,

Chehade *et al.* [CHE 08a] in 2008 paired an NSGA-II-type multi-objective genetic algorithm with a simulation model of a print shop to determine the best queue management policies for the simultaneous optimization of these criteria.

In Chapter 3, we will present an example that involves the coupling of simulation with an optimization, to solve a buffer-sizing problem.

Chapter 3

Design and Layout

3.1. Introduction

Decisions concerning the management of industrial systems may be taken in the long, medium or short term depending on their nature and the process under consideration. Designing or redesigning a system is a long-term problem. This chapter focuses on two large families of problems: the design of production lines and the design or physical layout of production facilities.

A production line design problem may be divided into different sub-problems and different constraints, as shown in Figure 3.1. The sub-problems include equipment selection, balancing and sizing of production lines, sizing of buffers, transport systems (conveyors), physical layout and grouping of production units. These different problems are interrelated. For example the selection of equipment will set conditions for balancing and sizing of the process, which will in turn assign workloads to stations. The physical layout of the process will be prepared accordingly. We should, therefore, address these problems in a hierarchical manner.

There are two types of production processes: continuous processes (as in the chemical or oil industries) and discrete processes (as in the electronics and automotive industries). This classification depends on the nature of the flow a system has to go through. Here, we are mainly interested in discrete production systems. These are classified into two categories. The first is known as *assembly* that involves a set of operations combining two or more components to form another component or product, such as in automobile assembly lines and food packaging lines. The second category invloves non-assembly operations, such as turning, milling, etc., which are known as machine operations.

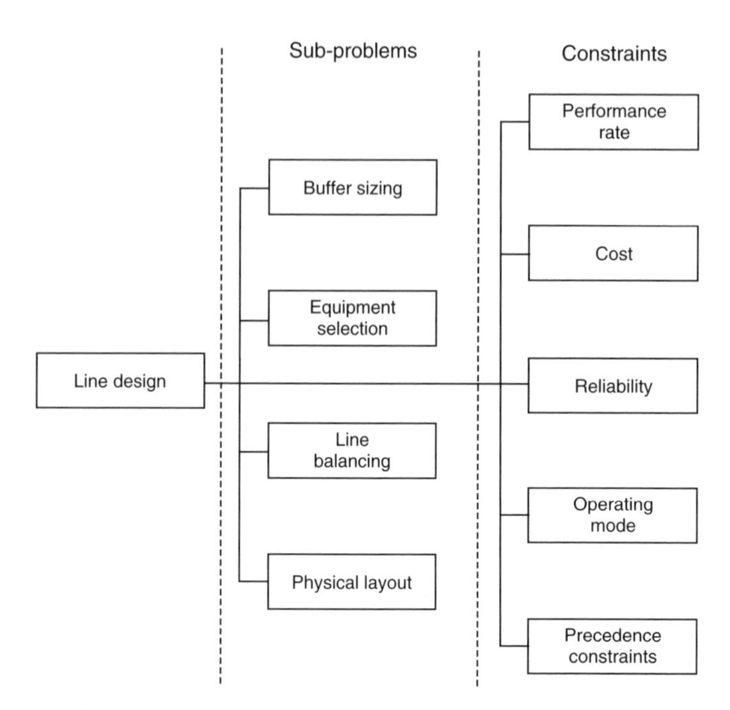

Figure 3.1. *The main problems and constraints of line design*

The type of a discrete production system varies not only according to the nature of the process upon which it relies, but also according to its production volume [PAP 93]. Discrete production system may be a mass production system, batch production system, or a small series production system. Mass production systems are generally highly automated. For small series production systems, automation is less but flexibility is greater. In fact, a production system must be capable of being adapted to the evolution of the market and present a certain degree of modularity.

In this chapter we will focus on all the problems faced in designing a production system. First, we present different types of production systems that are commonly used in the industrial world. Then, we describe different problems that occurred during the design of industrial systems. For each problem, we present different approaches and examples of their application.

3.2. The different types of production system

As already mentioned, in production systems we often distinguish between assembly and machining operations. A production process is comprised of machining operations and assembly.

DEFINITION 3.1.– *An operation is an elementary part of the total work of the production system.*

DEFINITION 3.2.– *A workstation consists of several machines. In each workstation, one or more operations are carried out on the product.*

DEFINITION 3.3.– *A buffer is a storage area between two workstations. It temporarily stores the products coming from the upstream station before they enter the downstream station.*

A production line consists of a set of workstations that may or may not be separated by buffers. A loading station is located at the start of the line to feed it with the necessary raw materials. An unloading station is located at the end of the line to receive the finished product, ready to be distributed to the customers. Figure 3.2 presents an example of a production line. The flow of products from one workstation to the next is driven by a transfer system (conveyor, moving belt, etc.). The production line may also use inspection stations (which may be fixed or move along the line) whose objective is to verify the smooth operation of the manufacturing process.

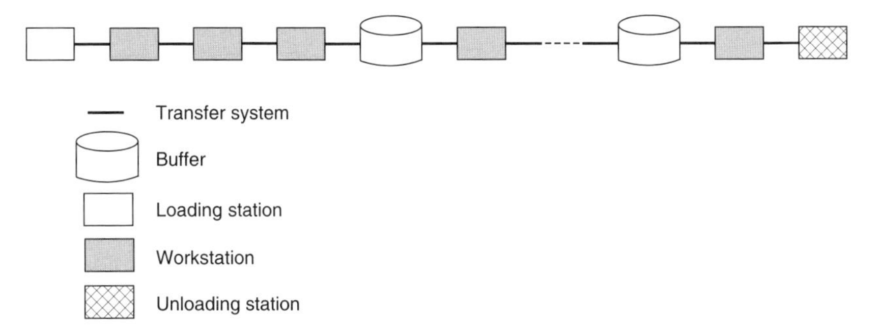

Figure 3.2. *A production line [MAK 05]*

DEFINITION 3.4.– *A flow shop system (Figure 3.3) is a production system in which the sequence of tasks (a task may represent one or more operations on the machines) is the same for all products (a single path). Such a system involves a sequence of workstations separated by intermediate storage areas.*

These types of systems are widely used today in the food and automotive industries. They may be classified according to the method of transferring products and the number of types of products. There are three methods of transferring products between workstations that determine which type of systems to be used:

– Synchronous: products are simultaneously transferred from one station to another. As a result, operating times are deterministic and equal for all the stations of the system, and therefore the system is completely balanced. These types of system are called transfer lines [MAK 05, MAK 08a, MAK 08b].

– Asynchronous: known as production lines, each product in this type of system is transferred independently of the others, thus yielding systems that are not necessarily balanced. Consequently, certain manual operations with variable cycles may be introduced. This type of system is often used for assembly or packing operations.

– Continuous: the products are transferred in a continuous manner at a constant speed. These systems set an upper limit to the time during which an operation is carried out. This upper limit is the time necessary to transfer a product by a conveyor between two points. This type of system required for this method is often used in assembly or packing processes.

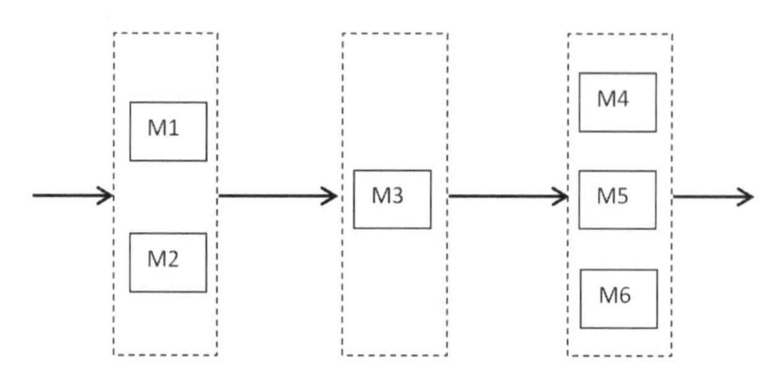

Figure 3.3. *An example of the structure of a flow shop-type system*

For products (their number and type) to flow through a system, a flow shop system is classified as follows:

– simple model: a single type of product is processed by the system;

– multiple model: two or more types of product are treated. These different types are not treated simultaneously, but in separate batches;

– mixed model: two or more types of products are processed simultaneously on the lines. The advantage of this model is the elimination of series-changing times, but its design and management are more difficult.

DEFINITION 3.5.– *Job shop systems (Figure 3.4) are also called multiple-path workshops. The structure of the workshop is the same as of a single-path workshop (flowshop) but its tasks may have different trajectories.*

DEFINITION 3.6.– *Open shop systems (Figure 3.5) are systems where each product is subjected to a set of operations on a sequence of workstations, but in a completely free order, i.e. at each workstation, where several machines may be installed, products may follow different paths depending on their type.*

DEFINITION 3.7.– *Assembly lines are production systems formed by a succession of workstations carrying out a set of tasks on a product [REK 06]. They have different types of operators (human or machine) that assemble the components of a product [MAK 05].*

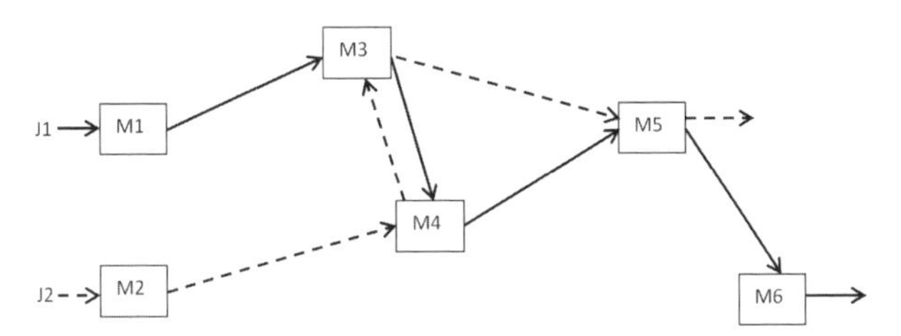

Figure 3.4. *An example of the structure of a job shop-type system*

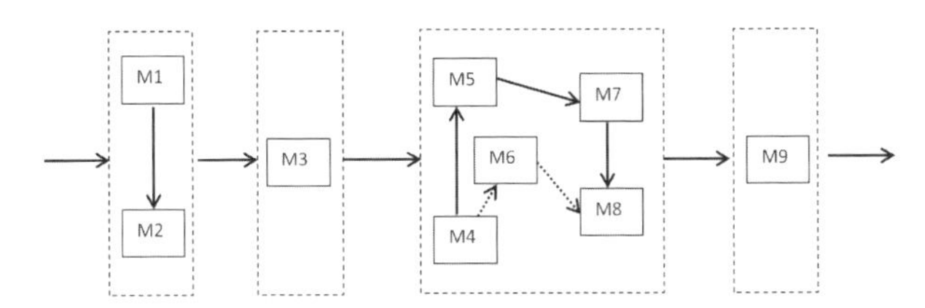

Figure 3.5. *An example of the structure of an open shop-type system*

In fact, the types of operators used in an assembly line determines the type of the line (manual assembly line, robotic line, etc.). The product assembled gradually takes its shape, starting with a base product to which other components are attached at the different stations visited.

3.3. Equipment selection

The first problem to be considered while designing a production system is the selection of equipment – which machines or technologies to choose in order to design a system that will meet several performance criteria. These criteria may belong to cost, productivity, or reliability of the system. If we choose one of these criteria to achieve optimization, the others may then become constraints, and we will have to solve a single-criterion problem. We may also choose to seek the best compromise between several criteria, leading to a multi-objective study. In the following section, we will give an overview of the evolution and the wealth of work on the problem of equipment selection. After this, we will describe some approaches to solving certain problems.

3.3.1. *General overview*

Graves and Whitney [GRA 79b] were among the first researchers to work on the problem of equipment selection. They proposed a linear programming model enabling the design of a system satisfying the constraints of the tasks with the lowest cost. The model is as follows [GRA 79b]:

$$Minimize \left[\sum_{i=1}^{N}\left(K_i.\delta_i + \sum_{j=1}^{M}(a_{ij}.X_{ij} + b_{ij}.X_{ij})\right)\right] \quad [3.1]$$

under the constraints:

$$\sum_{j=1}^{M}(t_{ij}.X_{ij} + l_{ij}.Y_{ij}) \leq U_i.\delta_i; \forall i = 1, \ldots, N \quad [3.2]$$

$$\sum_{i=1}^{N} X_{ij} = 1; \forall j = 1, \ldots, M \quad [3.3]$$

$$Y_{ij} \geq X_{ij} - X_{i,j-1}; \forall i = 1, \ldots, N; \forall j = 2, \ldots, M \quad [3.4]$$

$$X_{ij} \in \{0, 1\}; \forall i = 1, \ldots, N \quad [3.5]$$

$$X_{ij}, Y_{ij} \geq 0; \forall i = 1, \ldots, N; \forall j = 1, \ldots, M \quad [3.6]$$

– N: the number of assembly stations;

– M: the number of tasks;

– K_i: the annual cost of station i;

– U_i: the availability rate of station i;

– t_{ij}: the operating time of task j assigned to station i;

– l_{ij}: the loading/unloading time of task j assigned to station i;

– a_{ij}: the annual operating cost of task j assigned to station i;

– b_{ij}: the annual operating cost saved by assigning j-1 to station i;

– δ_i: a binary variable which equals 1 if station i is chosen and 0 otherwise;

– X_{ij}: the annual volume of task j assigned to station i; and

– Y_{ij}: a variable which specifies the portion of task j assigned to station i.

The objective function (equation [3.1]) represents the investment and operational costs as a function of the decision variables. The first constraint (equation [3.2]) ensures that the availability of each station is not exceeded. The second constraint (equation [3.3]) is the standard assignment constraint, which guarantees the complete allocation of the annual volume of each task. The third constraint, equation [3.4]), relates Y_{ij}, the load indicator, to X_{ij}, which is the allocation variable. Combining the constraint of equation [3.4] with that of equation [3.6], Y_{ij} equals zero only if $X_{i,j-1}$ is greater than or equal to X_{ij}. In other words, the quantity of tasks j assigned to station i equals 0 if station i carries out task $j-1$ as often as task j. The constraint of equation [3.4] represents the binary variable δ_i.

The most studied criterion in equipment selection problems is the minimization of the cost of the designed system. In 2003, Dolgui *et al.* [DOL 03] studied this problem with a single criterion: the minimization of the cost of the lines. The lines were considered as being formed of parallel operation blocks. The same objective was taken into consideration in the work of Makdessian *et al.* [MAK 08b] with certain technological constraints.

Another criterion that is often taken into consideration in equipment selection problems is the maximization of the performance rate, such as in the work of Koren *et al.* [KOR 99] or Chehade *et al.* [CHE 08b]. Other criteria are presented in the work of Cetinkaya [CET 07], who proposed a new approach concerning the selection of equipment for an automated line. These criteria include the customer service rate, an environmental criterion, the precisions of the system, the production strategy, etc.

3.3.2. *Equipment selection with considerations of reliability*

3.3.2.1. *Introduction to reliability optimization*

In order to ensure the security and control of production systems, their reliability is to be considered from their inception. The more reliable the production tool, the higher will be the level of security. Different methods exist to guarantee that a system has a high level of reliability: duplicate a strategic resource in a production line, put in place curative or preventative maintenance plans, etc.

In the scientific literature, problems that consist of attributing a reliability to a component (for example a machine) of a system to reach a minimal objective are called reliability allocation problems [KUO 01]. If we examine the number of machines to be installed in parallel (identical or not, but having the same function) in a workstation, then this is called redundancy allocation. Problems that combine both may be encountered, and are called combined redundancy and reliability allocation. The vast majority of these problems are NP-hard [CHE 92]. Reliability may also be considered as an objective to be maximized, with constraints of cost and volume. It may equally be expressed in the form of a constraint in a problem minimizing the cost of purchasing machinery. One may also encounter multi-objective optimization problems in which the maximization of reliability and the minimization of cost are considered together. An overview of the existing work in the domain of reliability optimization is given by A. Yalaoui [YAL 04].

In general, we may distinguish between two approaches:

– We seek to allocate a reliability to each component of the system to obtain a minimal overall reliability, without concern for technologies, components, or machines available in the market. This approach may be interesting at inception to provide recommendations for specifications, for example before examining the choice of equipment. We may also find ourselves in the situation where we must make customized machines, and such a study will allow us to specify the characteristics of the machines to be designed. The work of Elegbede *et al.* [ELE 03] gives an example approach in this case.

– We begin with the products available on the market, characterized, among other things, by their reliability, and we seek those which lead to a system that will adhere to a minimum reliability constraint. This approach corresponds to the problem of equipment selection itself.

3.3.2.2. *Design of a parallel-series system*

3.3.2.2.1. Definition of the system and models

Parallel-series systems are constituted by several workstations in series. Each workstation is equipped with one or more machines (not necessarily identical, but with the same functionalities) in parallel. There are several ways to connect machines in parallel. We may consider the use of a single machine, and when it breaks down another takes over (which is known as passive redundancy), or the use of all the machines at the same time (known as active redundancy). Here, we will consider the latter case.

Let us consider a series composed of s subsystems (or workstations). Each subsystem i $(1 \leq i \leq s)$ is composed of actively redundant components. We seek to determine, under a minimum reliability constraint $R_{\min}$, the number of machines to be put in parallel and to choose from them to minimize the total cost of the system. For each subsystem i, we know the number n_i of types of machine available in the market (in unlimited quantity). Each type of machine k is characterized by its reliability, written p_{ik}, and its unit cost, written c_{ik} $(1 \leq k \leq n_i)$. We write as $\Omega_i = \{p_{i1}, p_{i2}, \ldots, p_{in_i}\}$ the set of reliabilities of the machines available for subsystem i. We suppose that $p_{i1} < p_{i2} < p_{i3} < \cdots < p_{in_i-1} < p_{in_i}$ and $c_{i1} < c_{i2} < c_{i3} < \cdots < c_{in_i-1} < c_{in_i}$ $(1 \leq i \leq s)$. We also suppose that for each subsystem i, we have a lower bound l_i and an upper bound u_i on the number of machines to be put in parallel k_i.

The decision variables are then the levels of redundancy (the number of machines in parallel) k_i, and the reliability of the machines r_{ij} $(1 \leq i \leq s,\ 1 \leq j \leq k_i)$. The reliability of the system is expressed as:

$$R_S = \prod_{i=1}^{s} \left(1 - \prod_{j=1}^{k_i} (1 - r_{ij})\right) \qquad [3.7]$$

and the cost is:

$$C_S = \sum_{i=1}^{s} \sum_{j=1}^{k_i} f_i(r_{ij}) \qquad [3.8]$$

where f_i maps the cost c_{ik} to the reliability p_{ik} for $k = 1, \ldots, n_i$. The equipment selection problem for a parallel-series production system is then written as:

$$\min \sum_{i=1}^{s} \sum_{j=1}^{k_i} f_i(r_{ij}) \qquad [3.9]$$

$$\prod_{i=1}^{s} \left(1 - \prod_{j=1}^{k_i} (1 - r_{ij})\right) \geq R_{\min} \qquad [3.10]$$

$$l_i \leq k_i \leq u_i \; \forall i = 1, \ldots, s \qquad [3.11]$$

$$r_{ij} \in \Omega_i, \;\; \forall i = 1, \ldots, s, \; j = 1, \ldots, k_i \qquad [3.12]$$

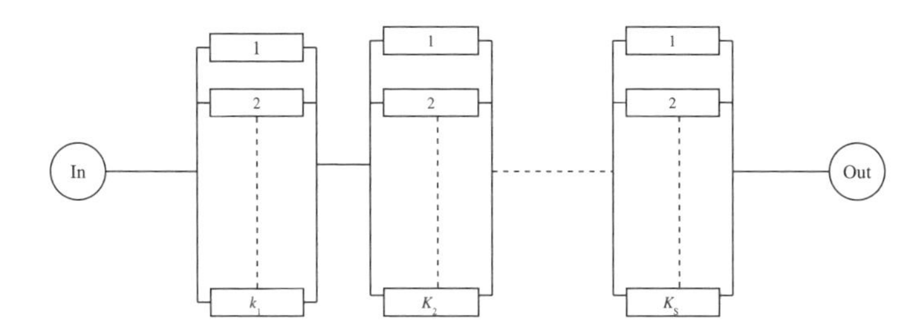

Figure 3.6. *A parallel-series system*

We may transform this mixed nonlinear programming problem into a nonlinear integer programming problem by introducing a new variable x_{ik} representing the number of machines of type k $(1 \leq k \leq n_i)$ used in the subsystem i $(i = 1, \ldots, s)$. We then have:

$$k_i = \sum_{k=1}^{n_i} x_{ik}, \;\; i = 1, \ldots, s \qquad [3.13]$$

The reliability of the system is:

$$R_S = \prod_{i=1}^{s} \left[1 - \prod_{k=1}^{n_i} (1 - p_{ik})^{x_{ik}} \right] \quad [3.14]$$

and the cost:

$$C_S = \sum_{i=1}^{s} \sum_{k=1}^{n_i} c_{ik} x_{ik} \quad [3.15]$$

Thus we obtain the following mathematical formulation, which is equivalent to the previous model:

$$\min \sum_{i=1}^{s} \sum_{k=1}^{n_i} x_{ik} c_{ik} \quad [3.16]$$

$$\prod_{i=1}^{s} \left[1 - \prod_{k=1}^{n_i} (1 - p_{ik})^{x_{ik}} \right] \geq R_{\min} \quad [3.17]$$

$$l_i \leq \sum_{k=1}^{n_i} x_{ik} \leq u_i \; \forall i = 1, \ldots, s \quad [3.18]$$

$$x_{ik} \in \mathbb{N}, \quad \forall k = 1, \ldots, n_i \; \forall i = 1, \ldots, s \quad [3.19]$$

3.3.2.2.2. Solution method

We consider a one-level problem that includes sub-problem i of an s-level system. We suppose that its reliability objective, written $R_{i,\min}$, is known. Thus we solve the following problem:

$$\min \sum_{k=1}^{n_i} x_{ik} c_{ik} \quad [3.20]$$

$$1 - \prod_{k=1}^{n_i} (1 - p_{ik})^{x_{ik}} \geq R_{i,\min} \quad [3.21]$$

$$l_i \leq \sum_{k=1}^{n_i} x_{ik} \leq u_i \quad [3.22]$$

$$x_{ik} \in \mathbb{N}, \quad \forall k = 1, 2, \ldots, n_i \quad [3.23]$$

The solution method is based on the transformation of the problem into a one-dimensional knapsack problem. For every type of machine k ($1 \leq k \leq n_i$), the quantity $v_{ik} = -\ln(1 - p_{ik})$ is defined. This lets us write the following model:

$$\min \sum_{k=1}^{n_i} x_{ik} c_{ik} \qquad [3.24]$$

$$-\ln(1 - R_{i,\min}) \leq \sum_{k=1}^{n_i} x_{ik} v_{ik} \qquad [3.25]$$

$$l_i \leq \sum_{k=1}^{n_i} x_{ik} \leq u_i \qquad [3.26]$$

$$x_{ik} \in \mathbb{N}, \ \ \forall k = 1, 2, \ldots, n_i \qquad [3.27]$$

Writing $z_{ik} = u_i - x_{ik}$, we obtain the following model:

$$\max \sum_{k=1}^{n_i} z_{ik} c_{ik} \qquad [3.28]$$

$$\ln(1 - R_{i,\min}) + u_i \sum_{k=1}^{n_i} v_{ik} \geq \sum_{k=1}^{n_i} z_{ik} v_{ik} \qquad [3.29]$$

$$u_i(n_i - 1) \leq \sum_{k=1}^{n_i} z_{ik} \leq n_i u_i - l_i \qquad [3.30]$$

$$0 \leq z_{ik} \leq u_i \ \ \forall k = 1, 2, \ldots, n_i \qquad [3.31]$$

$$z_{ik} \in \mathbb{N} \qquad [3.32]$$

The above model is that of a one-dimensional knapsack problem. This family of problems is presented in the work of Martello and Toth [MAR 80]. In such a problem, we have several types of object of different values and volumes, as well as a bag of limited capacity. We choose the combination of objects that will maximize the total profit without exceeding the volume of the bag. Here, the maximum volume of the bag is $V_{i,\max} = \ln(1 - R_{i,\min}) + u_i \sum_{k=1}^{n_i} v_{ik}$, z_{ik} is the number of objects chosen of type k, c_{ik} represents the profit given by an object of type k and v_{ik} is its volume. The total number of objects must lie between a lower bound, written $b_i^- = u_i(n_i - 1)$, and an upper bound, written $b_i^+ = n_i u_i - l_i$. Moreover, the number z_{ik} of objects of type k to be put in the bag must not exceed u_i. This comes from the fact that $x_{ik} = u_i - z_{ik} \geq 0$. Finally, we obtain:

$$\max \sum_{k=1}^{n_i} z_{ik} c_{ik} \qquad [3.33]$$

$$V_{i,\max} \geq \sum_{k=1}^{n_i} z_{ik} v_{ik} \qquad [3.34]$$

$$b_i^- \leq \sum_{k=1}^{n_i} z_{ik} \leq b_i^+ \qquad [3.35]$$

$$0 \leq z_{ik} \leq u_i \;\; \forall k = 1, 2, \ldots, n_i \qquad [3.36]$$

$$z_{ik} \in \mathbb{N} \qquad [3.37]$$

The most well-known method, and the most adapted to solving knapsack problems, is dynamic programming [BEL 57]. It lends itself particularly well here to the problem considered by its recursive formulation.

Let $\Phi_i(k, n, V)$ be the profit function, which represents the maximum profit given by exactly n objects among those of types 1 to k such that the total volume is exactly V. We also define $z_{ik}^*(n, V)$ as the optimal number of objects of type k to be taken to obtain the maximum profit $\Phi_i(k, n, V)$. We then obtain the following recursive formulation, where $n = \sum_{l=1}^{k} z_{il}$ is the cumulative number of objects whose type is less than or equal to k ($1 \leq k \leq n_i$):

– initial conditions:

$$\begin{cases} \Phi_i(0,0,0) = 0 \\ \Phi_i(0,n,V) = -\infty \;\; \forall\, n > 0, \;\; \forall\, V > 0 \\ \Phi_i(k,0,V) = -\infty \;\; \forall\, V > 0, \;\; 1 \leq k \leq n_i \end{cases} \qquad [3.38]$$

– recurrence formula:

$$\begin{cases} \Phi_i(k,n,V) = \max\limits_{0 \leq z \leq \min(n,u_i)} \{\Phi_i(k-1, n-z, V - z v_{ik}) + c_{ik} z\} \\ z_{ik}^*(n,V) = \arg \max\limits_{0 \leq z \leq \min(n,u_i)} \{\Phi_i(k-1, n-z, V - z v_{ik}) + c_{ik} z\} \end{cases} \qquad [3.39]$$

The optimal solution corresponds to:

$$\max_{b_i^- \leq n \leq b_i^+, \;\; 0 \leq V \leq V_{\max}} \Phi_i(n_i, n, V) \qquad [3.40]$$

Solving the problem then consists of n_i sequential decisions. At each decision level k ($1 \leq k \leq n_i$), several states are possible for the system. Each state is characterized by two values: the cumulative number n of objects of type less than or equal to k and the volume V, which may be any real number less than $V_{i,\max} = \ln(1 - R_{i,\min}) + u_i \sum_{k=1}^{n_i} v_{ik}$. This renders the enumeration of every pair (n, V) impossible. To solve this problem, it is necessary to consider only the positive integer values below $V_{i,\max}$.

However, this implies a loss of precision and therefore the optimality of the solution. In order to conserve the greatest possible precision, an integer parameter L is introduced, and all of the integer values of $V_{i,\max} \times L$ are considered. The greater the value of L, the greater the precision of the results. The recursive formulation then becomes:

$$\begin{cases} \Phi_i(k,n,V) = \max_{0 \leq z \leq \min(n,u_i)} \{\Phi_i(k-1, n-z, V - z\lfloor v_{ik} L \rfloor) + c_{ik} z\} \\ z^{*L}(n,V) = \arg \max_{0 \leq z \leq \min(n,u_i)} \{\Phi_i(k-1, n-z, V - z\lfloor v_{ik} L \rfloor) + c_{ik} z\} \end{cases}$$

[3.41]

where V is every integer such that $0 < V \leq V_{i,\max} \times L$. z_{ik}^{*L} $(1 \leq k \leq n_i)$ is the optimal number of k-type objects used in sub-problem i giving the maximum profit $\Phi_i(n_i, n, \lfloor L \sum_{k=1}^{n_i} z_{ik}^{*L} v_{ik} \rfloor)$ for a given L. The corresponding solution for sub-problem i in the selection of machines is obtained from the following relation:

$$x_{ik}^{*L} = u_i - z_{ik}^{*L} \ \forall \ k = 1, \ldots, n_i \qquad [3.42]$$

The corresponding reliability R_i^{*L} and the cost C_i^{*L} are:

$$R_i^{*L} = 1 - \prod_{k=1}^{n_i} (1 - p_{ik})^{x_{ik}^{*L}} \qquad [3.43]$$

$$C_i^{*L} = \sum_{k=1}^{n_i} c_{ik} u_i - \max_{u_i(n_i - 1) \leq n \leq n_i u_i - l_i} \Phi_i(n_i, n, \lfloor L \sum_{k=1}^{n_i} z_{ik} v_{ik} \rfloor) \qquad [3.44]$$

From the assumption about the availability of machines on the market, there are only a finite number n_i of types of machine able to perform the same task for each subsystem i. This implies that the reliability of the subsystems may also take only a finite number of values between zero and one. A lower bound, written $\underline{R_i}$, and an upper bound, written $\overline{R_i}$, on the reliability of each subsystem i are given by:

$$\overline{R_i} = 1 - (1 - p_{in_i})^{u_i} \qquad [3.45]$$

$$\underline{R_i} = \frac{R_{\min}}{\prod_{j=1, j \neq i}^{s} \overline{R_j}} \qquad [3.46]$$

Thus $\underline{R_i} \leq R_{i,\min} \leq \overline{R_i}$ and the equivalent limits on the volume $V_{i,\max}$, written $\underline{V_i}$ and $\overline{V_i}$, are:

$$\overline{V_i} = \ln(1 - \underline{R_i}) + u_i \sum_{k=1}^{n_i} v_{ik} \qquad [3.47]$$

$$\underline{V_i} = \ln(1 - \overline{R_i}) + u_i \sum_{k=1}^{n_i} v_{ik} \qquad [3.48]$$

The mathematical formulation of the total problem, with the reliabilities R_i ($i = 1, \dots, s$) of the subsystems as decision variables, is the following:

$$\min \sum_{i=1}^{s} C_i(R_i) \qquad [3.49]$$

$$\prod_{i=1}^{s} R_i \geq R_{\min} \qquad [3.50]$$

$$R_i \in \{R_{i,1}, \dots, R_{i,N_i}\} \; \forall\, i = 1, \dots, s \qquad [3.51]$$

where $C_i(R_{i,j}) = C_{i,j}$ is the cost function. $\{R_{i,1}, \dots, R_{i,N_i}\}$ is the set of solutions for each subsystem i such that the corresponding final volume lies between $\lfloor \underline{V_i} \times L \rfloor$ and $\lfloor \overline{V_i} \times L \rfloor$. An equivalent problem may be obtained by introducing the variables $V_i = -\ln(R_i)$ ($i = 1, \dots, s$). Let us also write $V_{\max} = -\ln(R_{\min})$ for the maximum volume of the bag. Let $b_i(V_i) = -C_i(R_i)$ be the profit given by an object of volume V_i. We then obtain the following equivalent formula:

$$\max \sum_{i=1}^{s} b_i(V_i) \qquad [3.52]$$

$$\sum_{i=1}^{s} V_i \leq V_{\max} \qquad [3.53]$$

$$V_i \in \{V_{i,1}, \dots, V_{i,N_i}\} \; \forall\, i = 1, \dots, s \qquad [3.54]$$

This model corresponds to a one-dimensional knapsack problem, in which we must put exactly s objects, one of each type. For each type i ($i = 1, \dots, s$), we choose one and only one object in the set N_i of the different objects, characterized by their volume $V_{i,j} = -\ln(R_{i,j})$ and their profit $b_i(V_{i,j}) = -C_i(R_{i,j})$.

To solve this problem, dynamic programming is used again. The profit function $\psi(i, \omega)$ is the maximum profit given by the objects of types 1 to i ($i = 1, \dots, s$), such that the total volume used by these objects does not exceed ω. An integer L is introduced so that we are able to list all the attainable volumes with varying degrees of precision. The recurrent formulation of dynamic programming is then:

– initial conditions:

$$\begin{cases} \psi(0, \omega) = 0, \; \forall\, \omega \geq 0 \\ \psi(i, 0) = 0, \; \forall\, i \geq 0 \end{cases} \qquad [3.55]$$

– recurrence formulation:

$$\begin{cases} \psi(i, \omega) = \max\limits_{j \in \{1,2,\dots,N_i\}} \{\psi(i-1, \omega - V_{ij}^L) + b_i(V_{ij}^L)\} \\ j_i^{*L}(\omega) = \arg \max\limits_{j \in \{1,2,\dots,N_i\}} \{\psi(i-1, \omega - V_{ij}^L) + b_i(V_{ij}^L)\} \end{cases} \qquad [3.56]$$

The optimal solution is given, for a fixed L, by $\psi(s, V_{\max}L)$. The equivalent solution of the equipment selection problem for the parallel-series production system is obtained with the following relations:

$$x_{ik}^{*L} = u_i - z_{ik}^{*L} \ \forall \ i = 1, \ldots, s \text{ and } \forall \ k = 1, \ldots, n_i \qquad [3.57]$$

$$R_i^{*L} = 1 - \prod_{k=1}^{n_i} (1 - p_{ik})^{x_{ik}^{*L}} \ \ \forall \ i = 1, \ldots, s \qquad [3.58]$$

$$R^{*L} = \prod_{i=1}^{s} R_i^{*L} \qquad [3.59]$$

$$C^{*L} = -\psi(s, V_{\max}L) \qquad [3.60]$$

where R^{*L} is the final reliability of the system and C^{*L} is the corresponding minimum cost; both are functions of the given precision parameter L. This method, called the YCC algorithm, may be summarized as follows:

– step 1: for each subsystem $i = 1, \ldots, s$, calculate bounds $\underline{R_i}$ and $\overline{R_i}$;

– step 2: seek the N_i attainable solutions $R_{i,j}$ so that $\underline{R_i} \leq R_{i,j} \leq \overline{R_i}\ i = 1, \ldots, s$, by solving each subsystem i using dynamic programming; and

– step 3: solve the total problem by dynamic programming and keep the solution for which the reliability $R_s \geq R_{\min}$ has the lowest cost.

As the value of L increases, the algorithm converges toward the optimal solution.

3.3.2.2.3. An example of an application

Let us, first of all, consider the subsystem composed of actively redundant components. We suppose that four types of component are available in the market ($n_1 = 4$). The bounds on the number of components are $l_1 = 1$ and $u_1 = 2$. The characteristics of these four components are shown in Table 3.1.

k	p_{1k}	c_{1k}
1	0.560	4.480
2	0.680	12.330
3	0.800	28.310
4	0.880	45.980

Table 3.1. *The data for example 1*

The volume associated with a component of type k is obtained with $v_{1k} = -\ln(1 - p_{1k})$. We then obtain the following vector:

$$(v_{11}, v_{12}, v_{13}, v_{14}) = (0.810,\ 1.138,\ 1.622,\ 2.157) \qquad [3.61]$$

We consider a reliability objective of $R_{1\,\min} = 0.90$, which gives a maximum volume for the bag of $V_{1\,\max} \simeq 9.1514$. The model of the knapsack problem, with the parameter L fixed at 100, is as follows:

$$\max(4.48z_{11} + 12.33z_{12} + 28.31z_{13} + 45.98z_{14}) \quad [3.62]$$

$$915 \geq 81z_{11} + 113z_{12} + 162z_{13} + 215z_{14} \quad [3.63]$$

$$6 \leq \sum_{k=1}^{4} z_{1k} \leq 7 \quad [3.64]$$

$$0 \leq z_{1k} \leq 2 \quad k = 1, 2, 3, 4 \quad [3.65]$$

For $k = 1$, n may take the values 0, 1, or 2. Among all of the volumes V between 0 and 915, only three values (0, 81, and 162) are possible. The maximum profit for each of them is shown in Table 3.2.

$n\backslash V$	0	81	162
0	0	$-\infty$	$-\infty$
1	$-\infty$	4.48	$-\infty$
2	$-\infty$	$-\infty$	8.96

Table 3.2. *The states for $k = 1$*

Regarding the second step ($k = 2$), n may take only the values 2, 3 or 4. There are only six values possible for the volume (Table 3.3). For example, if $z_{11} = 2$ and $z_{12} = 2$, then the volume is $2 \times 81 + 2 \times 113 = 388$ and the corresponding maximum profit is $2 \times 4.48 + 2 \times 12.33 = 33.62$.

$n\backslash V$	162	194	226	275	307	388
2	8.96	16.81	24.60	$-\infty$	$-\infty$	$-\infty$
3	$-\infty$	$-\infty$	$-\infty$	21.29	29.14	$-\infty$
4	$-\infty$	$-\infty$	$-\infty$	$-\infty$	$-\infty$	33.62

Table 3.3. *The states for $k = 2$*

The states for the following steps are given in Table 3.4. There are seven possible values for the volume of step $k = 4$ that are 712, 765, 814, 818, 846, 876 and 899 with costs 90.24, 107.91, 123.89, 125.58, 131.74, 141.56 and 149.41, respectively. All these states are such that $n = 6$. There are no possible solutions for $n = 7$ at step 4, as all of the corresponding volumes exceed 915, the maximum permitted volume.

The maximum profit (for $L = 100$) is 149.41 for a volume of 899. It is obvious that $(z_{11}^{*100}, z_{12}^{*100}, z_{13}^{*100}, z_{14}^{*100}) = (1, 2, 1, 2)$. With the change of variable, we obtain

$(x_{11}^{*100}, x_{12}^{*100}, x_{13}^{*100}, x_{14}^{*100}) = (1,0,1,0)$. The reliability is then $R_1^{*100} = 1 - (1 - 0.56)^1(1 - 0.68)^0(1 - 0.80)^1(1 - 0.88)^0 = 0.912$ and the corresponding cost is $C_1^{*L} = 2 \times (4.48 + 12.33 + 28.31 + 45.98) - 149.41 = 32.79$.

$n\backslash V$	388	437	469	486	518	550	599	631	712
4	33.62	49.6	57.45	65.58	73.43	81.22	$-\infty$	$-\infty$	$-\infty$
5	$-\infty$	$-\infty$	$-\infty$	$-\infty$	$-\infty$	61.93	77.91	85.76	$-\infty$
6	$-\infty$	$-\infty$	$-\infty$	$-\infty$	$-\infty$	$-\infty$	$-\infty$	$-\infty$	90.24

Table 3.4. *The states for $k = 3$*

Let us now consider the design of a system that consists of two subsystems in a series. The first subsystem has the same characteristics as those given in Table 3.1. For the second subsystem, we suppose that two types of different components are available in the market ($n_2 = 2$), whose characteristics are indicated in Table 3.5. The bounds on the number of components in active redundancy are $l_2 = 1$ and $u_2 = 3$. The reliability objective for the system is $R_{\min} = 0.88$. We obtain $v_{21} = 0.982$ and $v_{22} = 2.461$. The solution of the two subsystems gives $N_1 = 8$ possible solutions for sub-problem 1 and $N_2 = 6$ for the second sub-problem (Table 3.6).

k	p_{2k}	c_{2k}
1	0.630	12.71
2	0.910	39.79

Table 3.5. *The data for the second part of example 1*

k	R_{1k}	C_{1k}	R_{2k}	C_{2k}
1	0.9856	91.96	0.9993	119.37
2	0.9771	74.29	0.9970	92.29
3	0.9629	58.31	0.9880	65.21
4	0.9609	56.62	0.9473	38.13
5	0.9485	50.46	0.9927	79.58
6	0.9366	40.64	0.9680	52.50
7	0.9121	32.79		
8	0.8976	24.66		

Table 3.6. *The possible solutions to the sub-problems*

Applying dynamic programming to solve the total problem, with $L = 100$, we obtain the following allocation for the reliabilities of the two subsystems: $R_1^{*100} = 0.936$ and $R_2^{*100} = 0.9473$, which gives $R^{*100} = 0.8872$ for a total cost of $C^{*100} = 78.77$.

The final configuration chosen for the system is shown in Figure 3.7.

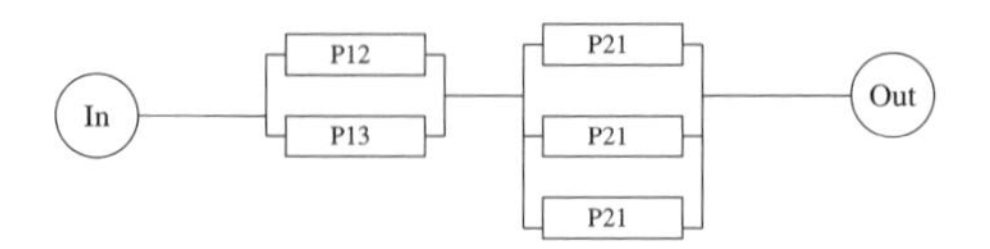

Figure 3.7. *A parallel-series system*

For this example, this solution cannot be improved for all $L > 100$; it is the optimal solution to the problem. This is not always the case, and sometimes the solution can be improved with a larger value of the parameter L, but the greater this value, the greater the run time.

3.4. Line balancing

Line balancing problems consist of allocating a workload (tasks) to each station in a line. This allocation is done so as to:

– minimize the unproductive time;

– minimize the number of workstations;

– maximize the output of the line;

– equally distribute the tasks to the stations; and

– avoid the violation of technical constraints, precedence constraints, etc.

A well-balanced line avoids the accumulation of outstanding tasks. The less space there is reserved to store stock, the smaller the cost of storage and the shorter the production time will be.

An inventory of the different criteria studied in the scientific literature was presented by Ghosh and Gagnon [GHO 89]:

– minimization of the number of stations with a given cycle time;

– minimization of the cycle time with a given number of stations;

– minimization of the idle time along the entire line;

– minimization of the delay;

– minimization of the line length;

– minimization of the exit time of the last vehicle;

– minimization of the probability that one or more stations exceed the cycle time;

– minimization of the labor cost per unit; and

– maximization of the profit.

Several constraints may also be taken into consideration in the solution of line balancing problems, such as [BOU 03]:

– scrolling time;

– precedence between operations;

– the distances required by the operations;

– incompatibility between operations;

– mandatory/forbidden allocations;

– the height or location of stations and operations;

– the grouping of operations; and

– the movement of products.

3.4.1. *The classification of line balancing problems*

Assembly line balancing problems are divided into two models [BAY 86]:

– the simple assembly line balancing (SALB) model, which covers a single type of product [BRO 06]. The lines have characteristics such as fixed cycle times, deterministic operation times, no allocation restrictions and a series speed; and

– the general assembly line balancing (GALB) model, which covers more problems than SALB such as mixed-model balancing, parallel stations, U-shaped lines and lines with stochastic operation times [LEV 06].

3.4.1.1. *The simple assembly line balancing model (SALB)*

Different types of problem exist for the SALB model:

– SALB-1: this problem consists of minimizing the number of stations in the line for a given cycle time;

– SALB-2: this problem consists of minimizing the cycle time for a fixed number of stations;

– SALB-E: this problem consists of maximizing the efficiency of the line with a given number of stations; and

– SALB-F: this problem consists of studying the feasibility of the operation to be assigned for a given cycle time and a fixed number of stations.

3.4.1.2. *The general assembly line balancing model (GALB)*

The general model of assembly line balancing includes several problems:

– The problem of cost-oriented balancing (CALB): the objective here no longer consists of minimizing the number of stations, but rather the cost generated by the configuration of the line.

– The problem of balancing U-shaped lines (U-shaped assembly line balancing – UALB): this problem considers packaging lines where the stations are placed one after another in a U shape. The operators may therefore work on both extremities of the line (entrance and exit). As a result, modified precedence constraints may be taken into consideration. Like SALB problems, several types of UALB problems also exist: UALB-F (the feasibility of operations), UALB-1 (minimizing the number of stations), UALB-2 (minimizing the cycle time) and UALB-E (maximizing the efficiency of the line).

– Mixed-model assembly line balancing (MALBP): this type of problem concerns lines with different models of the base product and a different sequence for each model. It consists of assigning tasks to stations taking into consideration the task time that varies according to the different models of the product. For this type of problem, we distinguish between MALBP-1, MALBP-2 and MALBP-E.

– Robotic assembly line balancing (RALB): this problem was introduced in 1991 by Rubinovitz and Bukchin [RUB 91] and consists of studying the balancing of robotic lines.

3.4.2. *Solution methods*

In this section, we present some solutions applied to the most well-studied type of balancing problem: SALB.

3.4.2.1. *Exact methods*

Several exact solution methods have been proposed for assembly line balancing problems. These methods are generally based on two approaches – dynamic programming (an approach that may be station-oriented or operation-oriented) and the branch-and-bound procedure. The latter may involve several methods:

– The FABLE method establishes a pattern of operation-oriented branching, i.e. a branch corresponds to the allocation of a single operation at once. This method was proposed by Johnson [JOH 88].

– The EUREKA method: the search tree is station-oriented, i.e. each branch corresponds to the total workload of a station. Also known as the lower bound method, EUREKA is an iterative process, which attempts to find a solution that does not exceed the best theoretical value of the number of stations, i.e. the value of the lower bound. The EUREKA method was introduced by Hoffmann [HOF 92].

– The SALOME method: the branch pattern is station-oriented. The principle of this method is to replace the iterative process of EUREKA that constructs several trees from one, only developing a single one. The SALOME method was proposed by Scholl and Klein [SCH 97].

3.4.2.2. *Approximate methods*

The approximate methods applied to solve assembly line balancing problems may be either heuristic or metaheuristic. The main proposed heuristics are:

– Ranked positional weight (RPW): this is a heuristic proposed by Kilbridge and Wester [KIL 61] based on a simple decision rule (a single solution is constructed, i.e. a single-pass heuristic).

– Computer method of sequencing operations for assembly lines (COMSOAL): this heuristic [ARC 66] uses a random choice, generating multiple solutions and keeping the best (a multi-pass heuristic). The aim is to load the stations as much as possible, giving priority to those located at the start of the line. The operations to be allocated are randomly chosen from those for which all the preceding operations have been performed.

Regarding metaheuristics, several methods have been proposed, such as those based on genetic algorithms, ant colonies and tabu search, among others.

3.4.3. *Literature review*

Assembly line balancing problems have been discussed in a large number of works, such as those of Bukchin and Rubinovitz [BUK 03], or Boysen *et al.* [BOY 06]. In 2000, Bukchin and Tzur [BUK 00] studied the problem of flexible assembly line balancing with a single type of product and deterministic operation times, seeking to minimize the equipment cost.

For more details on line balancing problems, solution and analysis techniques, as well as example solutions, the reader may consult the work of Dolgui and Proth [DOL 06a, DOL 06b]. This work brings together the more recent concepts and techniques for production line balancing.

3.4.4. *Example*

In its workstation, a company wishes to carry out the assembly of a product P. This assembly involves the completion of a number of operations, which we will label A to N. Figure 3.8 models the precedence constraints of these operations. The duration in minutes of each operation is presented in Table 3.7.

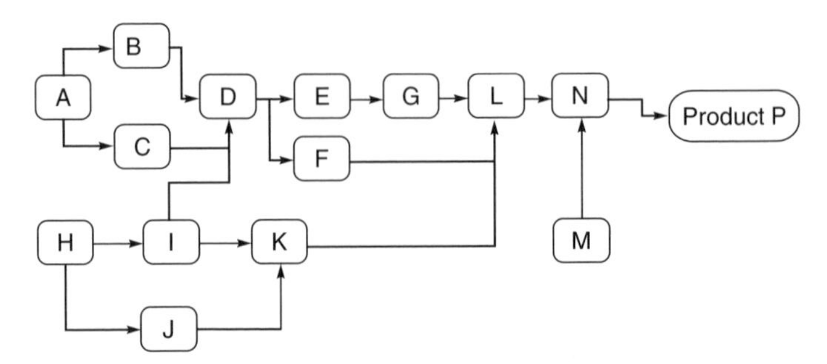

Figure 3.8. *The precedence graph of the operations in an assembly line*

Operation	A	B	C	D	E	F	G	H	I	J	K	L	M	N
Duration	4	5	2	6	1	5	6	1	5	4	2	5	1	5

Table 3.7. *The durations of the operations to be performed for the assembly of product P*

The first step is to determine the minimum cycle time to make one product. The operating time bottleneck equals six minutes. This corresponds to the two operations D and G. The minimum cycle time that we may have for this line is therefore equal to six minutes.

The following step consists of determining the minimum number of stations to attain this six-minute cycle time. The total duration of the operations to be carried out is 52 minutes. This means that we would need 52/6 (i.e. nine) workstations. To have a balanced line, each station would need to have an operational time of six minutes.

The last step consists of allocating the operations, from A to N, to the nine workstations in such a way as to comply with the six-minute cycle time and the precedence constraints between the different operations. An example of this allocation is given in Table 3.8.

From this allocation, we note that only stations 2 and 9 have operation times of five minutes. We therefore have two minutes of non-productivity over the 54 minutes of the set of operation times of the nine stations, giving a non-productivity of 3.70%.

3.5. The problem of buffer sizing

The buffers, or intermediate storage areas, have an important impact on the performance of a system. When they are well configured, they may contribute to the improvement of the total performance of the system. Their main purpose in the flow of the production line is to give each component of the system a certain degree of independence with respect to the others. In the absence of buffers, the only way for two connected stations to correctly function is their total synchronization. Also, the absence of buffers between two workstations may result in at least one of these two situations: the blockage of the first station or the failure of supply to the second. A station is blocked when its downstream buffer is full and occupied. A supply failure occurs when a station has finished the treatment of a product while its upstream buffer is empty and the upstream station is occupied or has no task to execute. In both situations, the performance rate of the system is affected.

Station	Operations performed	Operation time
1	A and C	6
2	B	5
3	D	6
4	E and F	6
5	G	6
6	H and I	6
7	J and K	6
8	L and M	6
9	N	5

Table 3.8. *The duration of the operations to be performed for the assembly of product P*

The determination of the capacity of the buffers, measured in terms of the number of products, must be based on the expected objectives of the system. A high-capacity buffer has positive effects on the performance of the system, but it may generate supplementary costs. These costs are especially represented by the effect on the product's cycle time and the area occupied. On the other hand, a low-capacity buffer lets us limit costs but may lead to a lower performance. It is therefore necessary to establish a compromise.

	Advantages	Disadvantages
High-capacity buffer	Avoids disruptions Allows the production needs to be satisfied Eliminates upstream blockages Eliminates upstream supply failures	Increases the cycle time Causes difficult to identify the sources of faults Increases costs Causes delays
Low-capacity buffer	Provides protection against statistical fluctuations Limits cycle times	Increases the risk of blockages and supply failures Increases disruptions

Table 3.9. *Advantages and disadvantages of the different buffer sizes*

3.5.1. *General overview*

There are many works that consider the problem of buffer sizing with two configurations: primal and dual [GER 00]. The aim of the primal configuration is to reduce the total capacity of the buffers while trying to achieve a minimum performance rate for the line. The dual configuration aims at maximizing the performance rate under a constraint on the total volume of the buffers in the system. Most works consider buffer sizing as a single-objective problem [ALT 02, HAM 06]. Certain works [ABD 06, DOL 02] take into consideration several criteria (the production rate, the cost of acquiring and installing the buffers and the storage cost), collecting them into a single-objective function with weighting factors. Other, more recent works focus on multi-objective optimization [CHE 07, CHE 09, CHE 11].

The solution of a buffer sizing problem may be obtained using analytic models such as Markov chains [BUZ 71, HEN 92], dynamic programming [CHO 87, JAF 89, YAM 98], or linear programming [ABD 06]. Performance evaluation models may also be used, either based on Petri nets [DSO 97] or with recourse to discrete-event simulation [LUT 98]. Finally, heuristics or metaheuristics are often applied to solve large problems, such as genetic algorithms [DOL 02] and ant colonies [CHE 11].

3.5.2. *Example of a multi-objective buffer sizing problem*

In this section, a multi-objective buffer sizing problem in an assembly line is considered. As well as buffers, the line consists of machines (which may fail) installed at different workstations in the line. Every pair of upstream and downstream stations is connected by a single buffer. As a result, the line uses N stations and N-1 buffers. The size of each buffer is bounded by a lower value (l_i) and an upper value (u_i).

The aim here is to determine the optimal capacities of the buffer by taking into considerations the technical characteristics of the stations, such as the processing time of the products (P), the mean time between failures (MTBF) and the mean time to repair (MTTR). The determination of the buffer capacities must optimize the considered systems according to two simultaneous objectives: the minimization of the total size of the buffer and the maximization of the performance rate of the line.

A mathematical model used for this system is given below. The two criteria C1 and C2 to be optimized are the total capacity of the buffers to be minimized and the performance rate of the line to be maximized, respectively. The decision variable Y_{ij} represents the size i of buffer j:

$$P_{DB} : Optimize(C1, C2) \quad [3.66]$$

$$C1 : Minimize \sum_{j=1}^{N-1} \sum_{i=1}^{B} Y_{ij}.b_{ij} \quad [3.67]$$

$$C2 : Maximize(E) \qquad [3.68]$$

under the constraints:

$$E = f(Y_{ij}) \qquad [3.69]$$

$$E_{lj}^B \leq E_{ij}^B \leq E_{uj}^B;\ \forall i = 1, \ldots, B;\ \forall j = 1, \ldots, N-1 \qquad [3.70]$$

$$\sum_{i=1}^{B} Y_{ij} = 1;\ \forall j = 1, \ldots, N-1 \qquad [3.71]$$

$$Y_{ij} \in \{0, 1\};\ \forall i = 1, \ldots, B;\ \forall j = 1, \ldots, N-1 \qquad [3.72]$$

– E: the performance rate of the line;

– N: the number of stations in the line;

– N-1: the number of intermediate buffers in the line;

– B: the number of possible sizes of a buffer, i.e. $B = u - l + 1$;

– Y_{ij}: a binary variable equal to 1 if size i is assigned to buffer j and 0 otherwise;

– b_{ij}: the size i of buffer j;

– E_{ij}^B: the output rate from buffer j having size i;

– E_{lj}^B: the lower bound for the output rate of buffer j;

– E_{uj}^B: the upper bound for the output rate of buffer j.

Constraint [3.69] means that the performance rate of the line varies as a function of the sizes allocated to buffers. Constraint [3.70] shows that the size of each buffer is bounded by a lower (l) and upper (u) value. Constraint [3.71] imposes that a single size be associated with each buffer. Finally, constraint [3.72] defines the binary decision variable.

3.5.3. *Example of the use of genetic algorithms*

The multi-objective genetic algorithm proposed to solve this problem is SPEA2, proposed by Zitzler *et al.* [ZIT 01]. This algorithm is based on the use of an archive that stores all the non-dominated solutions found during the iterations.

3.5.3.1. *Representation of the solutions*

The first step consists of creating an initial population P of size n. Each individual or chromosome represents one possible solution or configuration, determining the capacity of each buffer in the line. Each gene of the chromosome then represents a possible size of a buffer. The number of genes in the chromosome is thus equal to the number of buffers in the system. As a result, each gene may have a value V_i lying between the lower (l_i) and upper (u_i) bounds of the buffer capacities, as shown in [3.73]:

$$l_i \leq V_i \leq u_i \quad [3.73]$$

3.5.3.2. *Calculation of the objective function*

As we have already mentioned, the algorithm consists of creating an archive A of size m that contains the set of elite or non-dominated solutions in each iteration. The dominance of a solution over another by the SPEA2 algorithm is established by calculating the value of its objective function. The solutions whose objective function has a fitness value of less than 1 are non-dominated. The value of the objective function is calculated as follows.

The first step in calculating the adaptation value is performed using the set $P \cup A$ (just P for the first iteration). For each solution $x \in P \cup A$, we calculate the value $S(x)$ (equation [3.74]), which represents the number of solutions y dominated by x. $S(x)$ is known as the strength of a solution:

$$S(x) = |\{y/y \in P_t + A_t \wedge x \succ y\}| \quad [3.74]$$

where:

– $|.|$ represents the cardinality of a set;

– $+$ signifies the union of two sets; and

– x $\succ$ y signifies that x dominates y according to the Pareto dominance relation.

Then, a second value $R(x)$ is calculated according to equation [3.75], representing the set of solutions y that dominate x. If $R(i)$ equals 0, this means that the corresponding solution is non-dominated:

$$R(x) = \sum_{y \in P_t + A_t \wedge x \succ y} S(x) \quad [3.75]$$

In order to ensure diversity in the population, additional density information, based on the kth neighborhood method to differentiate between the individuals with the same $R(x)$ values, is adopted.

For each individual x, the distances that separate it from the other individuals y and from the archive are calculated and stored in a list. After sorting, element k gives a distance σ_x^k, where k is calculated according to equation [3.76]:

$$k = \sqrt{n + m} \quad [3.76]$$

where n is the size of the initial population and m is the size of the archive A.

The sought density $D(x)$ is calculated according to equation [3.77]:

$$D(x) = \frac{1}{\sigma_x^k + 2} \quad [3.77]$$

In the denominator, the number 2 is added to ensure that the value of the denominator is larger than 0 and that $D(x)$ is less than 1.

The value of the objective function may now be calculated as the sum of $R(x)$ and $D(x)$ (equation [3.78]):

$$F(x) = R(x) + D(x) \quad [3.78]$$

The non-dominated solutions are those with a value less than 1. All these solutions are thus transferred to the archive A.

3.5.3.3. *Selection of solutions for the archive*

When the non-dominated solutions are transferred to the archive A, three cases may occur: The first case does not require any particular procedure. It arises when the number of non-dominated solutions equals the size of the archive A. In the two other cases, we make a selection.

In the second case, the number of individuals exceeds the size of the archive. We then use the truncation operator: the individuals with the smallest distances from other individuals are eliminated.

On the other hand, if the number of non-dominated individuals is less than the size of the archive (the third case), the best dominated individuals of the previous population are selected.

3.5.3.4. *New population and stopping criterion*

The population of the following generation will be formed by a binary tournament of a selection of individuals from the population and the archive of the current generation. These selected individuals then undergo one-point crossover and mutation operations to achieve a greater diversity, thus forming the individuals of the new population.

The process is repeated until the stopping criterion, which in our case is the number of generations, is met. The overall structure of SPEA2 is presented in Algorithm 3.1.

3.5.4. *Example of the use of ant colony algorithms*

This section presents the application of a multi-objective ant colony optimization algorithm.

Algorithm 3.1 The structure of the SPEA2 algorithm [ZIT 01]

Generate an initial population P_0 and an empty archive $A_0 = \emptyset$
$t = 0$
Calculate $F(i)$ of the individuals in P_t and A_t (calculation of $S(i)$ then $R(i)$ and $D(i)$)
Copy the non-dominated individuals of P_t and A_t into A_{t+1}
if size $A_{t+1} > m$ **then**
  Reduce A_{t+1} with the truncation operator
end if
if size $A_{t+1} < m$ **then**
  Fill A_{t+1} with some dominated individuals from P_t and A_t
end if
if the stop criterion (i.e. the number of generations) is satisfied **then**
  $S = A_{t+1}$
else
  Choose individuals from P_t and A_t by binary tournament
  Apply crossovers and mutations
  Form P_{t+1}
  $t = t + 1$
end if

As mentioned in the previous chapter, the principle of ant colony optimization algorithms is based on the deposition of chemical substances (pheromones) by ants on their paths toward a source of food. The optimization algorithms are therefore based on pheromones, using matrices of quantities deposited between the different solutions, thus allowing the optimal solution to be sought. Given that multi-objective optimization takes into consideration several criteria, we will consider the total number of pheromone matrices to be equal to the number of objectives to be optimized. We therefore have two pheromone matrices for the multi-objective problem of buffer sizing presented above.

In the following section, we present the different steps of applying the Buffers-Multi-objective Ant Colony System (B-MOACS) algorithm to the problem being studied.

3.5.4.1. *Encoding*

The application of an ant colony algorithm to the problem of buffer sizing is based on the encoding presented in Figure 3.9. Each buffer i may have a capacity b_i which varies between a lower bound (l_i) and an upper bound (u_i).

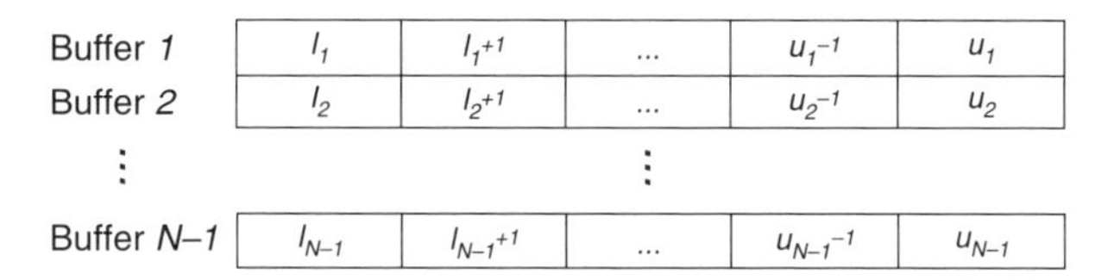

Figure 3.9. *A representation of the buffer encoding*

3.5.4.2. *Construction of the ant trails*

First, each ant is randomly placed on an initial point corresponding to the size of the first buffer. Then, each ant constructs its path by choosing the next points to visit. Each point on the path therefore represents the capacity of a buffer. The total number of points in each step corresponds to the different solution or to the capacities that each buffer may have. The choice of which points to visits based on the application of the state transition rule. An ant k at a point r chooses the next point s to visit according to equation [3.79]:

$$s = \begin{cases} argmax_{u \in J_k(r)} \left\{ \left[\sum_{o=1}^{O} w_o.\tau_{r,u}^o \right]^\alpha . [\eta_{r,u}]^\beta \right\} & \text{if } q \leq q_0 \\ S^* & \text{otherwise} \end{cases} \quad [3.79]$$

In equation [3.79], q is a randomly generated number between 0 and 1, q_0 is a parameter ($0 \leq q_0 \leq 1$) that determines the relative importance of exploitation against exploration. S^* is a random variable selected according to the probability given by equation [3.80]. O is the number of objectives considered. Elements o, $\tau_{r,s}^o$, and w_o are respectively the quantity of pheromones between points r and s and the weighting factor assigned to each objective. $\eta_{r,s}$ is a static value used as a heuristic of the innate desirability or the visibility of s from r. The calculation of the visibility is described in detail in the following section. The parameters α and β are used to give greater significance to either visibility or the quantity of pheromones. $J_k(r)$ is the set of points not visited yet by ant k:

$$S^* = \begin{cases} \dfrac{\left[\sum_{o=1}^{O} w_o.\tau_{r,s}^o \right]^\alpha . [\eta_{r,s}]^\beta}{\sum_{u \in J_k(r)} \left[\sum_{o=1}^{O} w_o.\tau_{r,u}^o \right]^\alpha . [\eta_{r,u}]^\beta} & \text{if } s \in J_k(r) \\ 0 & \text{otherwise} \end{cases} \quad [3.80]$$

3.5.4.3. *Calculation of the visibility*

A model based on an oriented graph is applied to calculate the visibility factor $\eta_{r,s}$ (Figure 3.10). Each capacity b_i of buffer i ($i = 1, \ldots, N$) is represented by a node. The set of nodes contains a source node U and a sink node V.

Each column in the graph represents the set of different capacities (l_i to u_i) of buffer i. The arcs connecting pairs of nodes serve to connect the capacities of consecutive buffers. The source node U has $(u_j - l_j + 1)$ arcs going to the nodes of the first buffer. Each arc linking the source node U to the capacity of the first buffer has distance d_i that is uniform across the arcs. Each arc between two nodes r and s has a distance equal to the cost of the two buffers i and i+1 which it links to the inverse of the maximum number of products leaving the system. This is presented in equation [3.81]:

$$d_{r,s} = (w_1.(b_i + b_{i+1}).U_{BC}) + \left(w_2.\frac{1}{E_{\max}}\right) \quad [3.81]$$

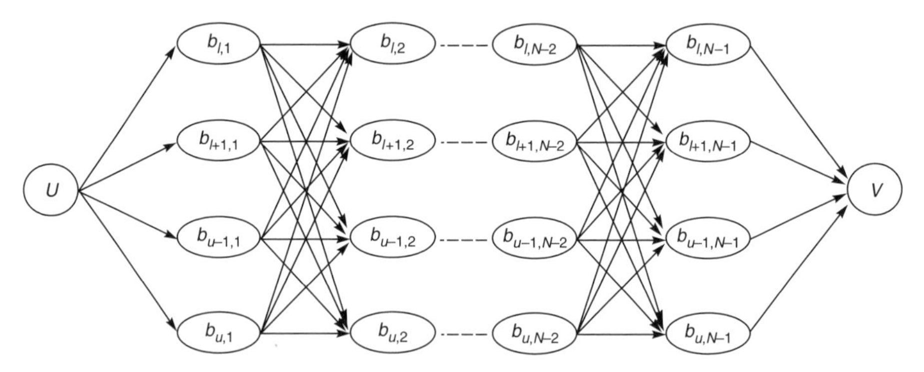

Figure 3.10. *A representation of a graph for the visibility calculation*

$E_{\max}$ is the maximum performance rate of the line when each buffer is assigned the maximum capacity. w_1 and w_2 are the weighting factors assigned to the first and the second objectives, respectively. We choose to give an identical significance to both objectives, which means that $w_1 = w_2 = 0.5$.

The use of these distances in an ant colony algorithm is useful for calculating the visibility factor $\eta_{r,s}$ that lets us choose the capacity of buffer i from buffer $i-1$. The visibility factor $\eta_{r,s}$ is calculated according to equation [3.82]. Its value is equal to the inverse of the distance of the arc linking two nodes, since we aim to reduce the cost of the configuration and maximize the performance rate of the line:

$$\eta_{r,s} = \frac{1}{d_{r,s}} \quad [3.82]$$

3.5.4.4. *Global and local updates of the pheromones*

In constructing its route, an ant changes the quantity of pheromones on the visited arcs by applying a local update function according to equation [3.83], where ρ is the pheromone evaporation rate ($0 \leq \rho \leq 1$) and τ_0 is the initial quantity of pheromones:

$$\tau^o_{r,s} = (1-\rho).\tau^o_{r,s} + \rho.\tau^o_0 \quad [3.83]$$

In our case, we consider τ_0 to be calculated according to equation [3.84] for the pheromone matrix of the first objective (cost), and according to equation [3.85] for the second objective (performance rate). We then choose to initialize the pheromones according to the higher capacities of the buffers:

$$\tau_0^1 = (U_{BC}. \sum_{i=1}^{N-1} b_{ui})^{-1} \qquad [3.84]$$

$$\tau_0^2 = E_{\max} \qquad [3.85]$$

Once all the ants have finished their route, the quantity of pheromones on each arc is modified according to a global update function as shown in equation [3.86]:

$$\tau_{r,s}^o = (1 - \rho).\tau_{r,s}^o + \rho.\Delta\tau_{r,s}^o \qquad [3.86]$$

$\Delta\tau_{r,s}^o$ is a factor that favors the non-dominated solutions. It is calculated according to equations [3.87] and [3.88]:

$$\Delta\tau_{r,s}^1 = \begin{cases} (C_{gb})^{-1} & \text{if arc } r, s \text{ belongs to a non-dominated solution} \\ 0 & \text{otherwise} \end{cases} \qquad [3.87]$$

$$\Delta\tau_{r,s}^2 = \begin{cases} E_{gb} & \text{if arc } r, s \text{ belongs to a non-dominated solution} \\ 0 & \text{otherwise} \end{cases} \qquad [3.88]$$

C_{gb} and E_{gb} represent the minimum cost and the maximum performance, respectively, found by the ants so far. With these updates to the pheromone matrices, the arcs that have higher quantities of pheromones will attract more ants in the following step. Algorithm 3.2 shows the total structure of the B-MOACS algorithm.

3.5.5. *Example of the use of simulation-based optimization*

Different factors limit the practical application of analytic methods (Markov processes, dynamic and linear programming, etc.) that cannot, for example model complex or large problems [LUT 98, PAP 01]. Thus the complex combinatorial nature of the problem of buffer sizing [HAR 99] requires the use of simulation, which seems more efficient at solving it [DOL 02, GER 00]. However, simulation models may in turn be limited by the number of possible scenarios. This leads to the pairing of optimization and simulation. The optimization model, generally based on efficient metaheuristics, is used to solve a problem that is combinatorial and stochastic in nature, whereas the simulation model is used to evaluate the performance of different configurations of the line.

Algorithm 3.2 The structure of the B-MOACS algorithm

Initialize the pheromones
Randomly place the k ants on the B capacities of the first buffer
for each buffer j = 2 to N-1 **do**
  Choose the next nodes to visit (buffer size j) based on equations [3.79] and [3.80]
end for
Return the ants to the departure points
Local updates to the pheromones according to equation [3.83]
for the k ants **do**
  Evaluate the solution proposed by the ant
end for
Retain the non-dominated solutions
Global update of the pheromones according to equation [3.86]
if the stop criterion is satified **then**
  Stop and exit the algorithm
else
  Construct new ant trails
end if

In the following section, we present an example of applying a simulation-based optimization technique to solve the buffer sizing problem presented previously [CHE 09]. Recall that the problem consists of sizing the buffers in a production line that is composed of seven stations with six intermediate buffers. A single machine is assigned to each workstation. Each machine is characterized by its processing time, its failure rate and its repair rate. Table 3.10 presents the different data for the machines [NAH 06].

N	$MTTR$	$MTBF$	T_i
1	450	820	40
2	760	5700	34
3	460	870	39
4	270	830	38
5	270	970	37
6	650	1900	40
7	320	1100	43

Table 3.10. *The input data for the studied problem*

The size of each buffer must lie between a lower value, 1 and an upper value, 20. The aim is then to determine the optimal capacity of each buffer in the line so as to optimize two criteria: the minimization of the total capacity of the buffers and the maximization of the number of products leaving the line.

To solve this problem, a simulation model is first developed with the ARENA® software. To optimize the performance of the model, the multi-objective genetic algorithm SPEA-II and the multi-objective ant colony optimization algorithm are applied. The simulation model and the optimization algorithms are used together to find the set of non-dominated solutions for the problem studied. We present a comparison between the two algorithms

3.5.5.1. *Simulation model*

The ARENA® simulation software is used to simulate the studied line. It is a program that is used to model and simulate different types of system. The simulation model proceeds as follows:

– creation of the flow of products;

– treatment of the products at the workstations; and

– discharge of the products.

The first step, which consists of creating the flow of products, is based on the use of the *Arrive* block of the ARENA® software. This block allows the flow to be generated by specifying the inter-arrival time of the products in the system. This time is set to one hour in our model. The block also allows the identification of the next destination of the products, which is station 1. Figure 3.11 shows the configuration of the *Arrive* block.

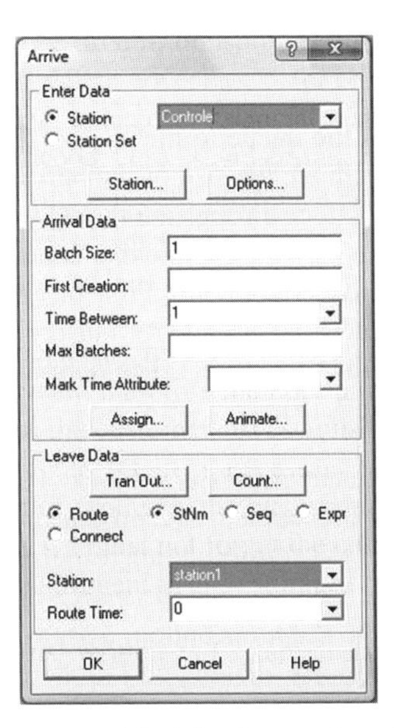

Figure 3.11. *The Arrive block for the creation of product flow*

The second step is the management of product treatment at the stations and the configuration of the buffers in the line.

To model a workstation with its upstream intermediate storage area (buffer), we have applied the model presented in Figure 3.12. In this model, the *Enter* block is used to receive the products at the station.

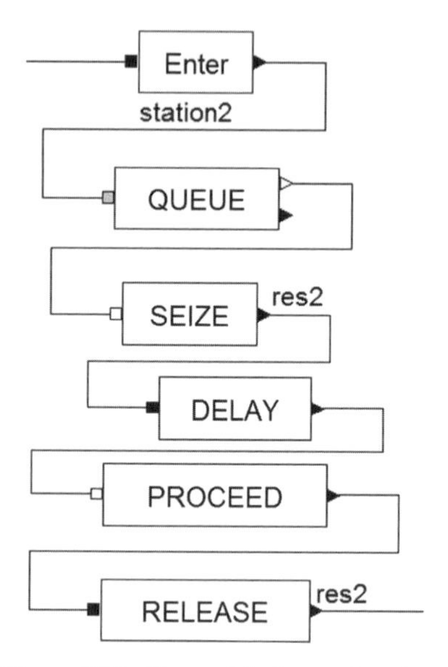

Figure 3.12. *The model of a workstation*

The *Queue* block models the queue to the station (Figure 3.13). Once the name of the queue (*Queue ID*) is defined, the size of the buffer must then be specified. The definition of the queue name is done through another *Queue* block from another library (Figure 3.14). In the latter, we may thus define the name and management rule of this queue. For this model, we have adopted the First In First Out (FIFO) policy. Since the size of the buffer is the decision variable for our problem, we have defined it as a variable (e.g. *capbuffer1* for the first buffer). The definition of these variables is done in the *Variables* block (Figure 3.15).

The block following the queue is the *Seize* block. This block models the fact that from this moment the station is occupied, as a product is being treated. In ARENA®, this translates to the seizure of the corresponding resource (e.g. *res2* for station 2). The configuration of this resource is done in the *Ressource* block as shown in Figure 3.16. In this block, by clicking on the *Resource* button, we may enter MTBF and the MTTR. This is done in the *Failures* part. By clicking on *Edit* here, a window appears allowing the values of two parameters *MTBF* and *MTTR* to be entered. Figure 3.17 shows an example of entering these values for station 2 (*MTBF* = 5700, *MTTR* = 760).

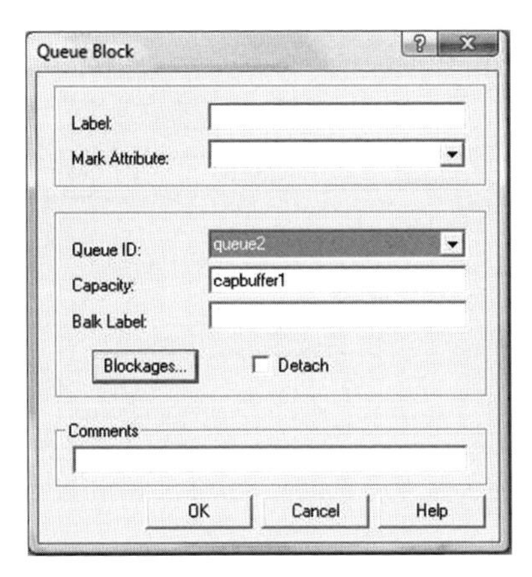

Figure 3.13. *The model of the queue*

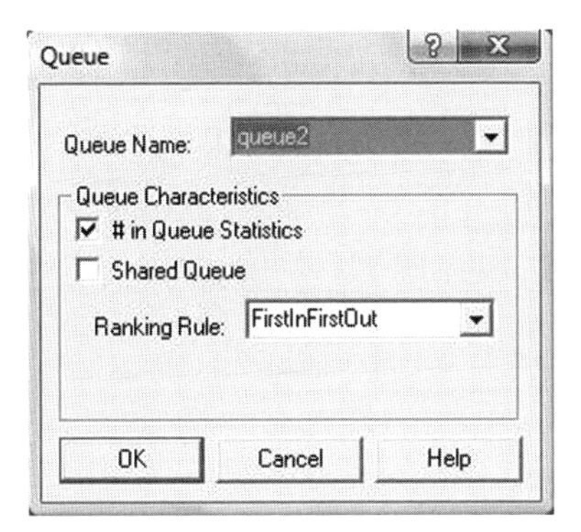

Figure 3.14. *The configuration of the queue*

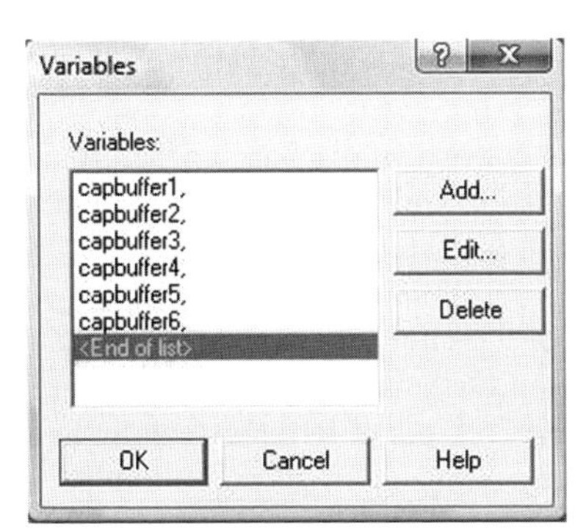

Figure 3.15. *The declaration of the model's variables*

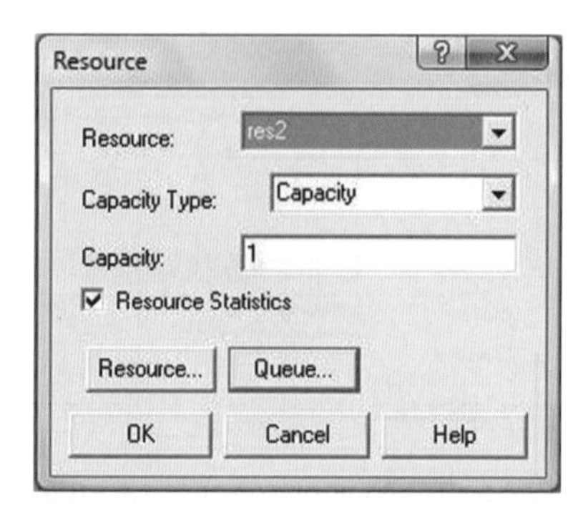

Figure 3.16. *The configuration of a workstation*

Figure 3.17. *Entering MTBF and MTTR values*

The next step occurs in the *Delay* and *Proceed* blocks, where we may enter the time corresponding to the processing time of the station. Once this time has elapsed, we may liberate the resource through the *Release* block.

The third and final step consists of discharging the products from the system using the *Depart* block. In this block, we may also set a counter that then lets us count the number of products finished in each replication.

The overall structure of the simulation model for the studied line is presented in Figure 3.18. The simulation time frame is fixed at a year for the simulation model with a transitional (*warm-up*) regime of 10 days. In order to limit the uncertainty in the evaluations, the latter is the average of 10 different simulations, known as replications.

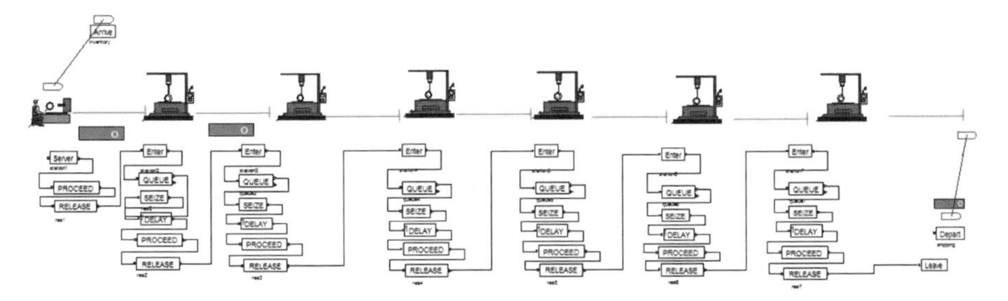

Figure 3.18. *The general structure of the simulation model*

3.5.5.2. *Optimization algorithms*

As stated in the previous section, two optimization algorithms are applied to solve the multi-objective buffer sizing problem under consideration. These two algorithms are MOACS (a multi-objective ant colony optimization algorithm) and SPEA-II (a multi-objective genetic algorithm). In the following, we present the configuration of these metaheuristics.

Several tests are performed in order to set the parameters of the algorithms developed to solve the considered problem: B-MOACS and SPEA2. In each test, we seek the best value for a single parameter while the others remain fixed. The final values are the result of a compromise between the quality of the final solutions and the required convergence time. The parameters of the B-MOACS and SPEA2 are presented in Tables 3.11 and 3.12, respectively.

Parameter	Value
k	40
ρ	0.7
q_0	0.7
α	1
β	2

Table 3.11. *The parameters of the MOACS algorithm*

Parameter	Value
Initial population	20
Number of generations	100
Size of the archive (A)	10
Number of crossover points	0.9
Probability of mutation	0.006

Table 3.12. *The parameters of the SPEA2 algorithm*

3.5.5.3. *The pairing of simulation and optimization*

This pairing is done using the technique of simulation-based optimization. A loop is generated, in which the optimization algorithm (in this example SPEA-II or MOACS) sends the values of the decision variables (the capacity of each buffer in our example) to the simulation model. The latter evaluates the performance of the current configuration according to the considered criteria: the number of output products and the total capacity of the buffers. Several round trips are necessary. The number is determined by the number of generations in the optimization algorithm.

This pairing is essentially managed by the Visual Basic for Applications (VBA) interface of the ARENA® software. At this interface, we begin by declaring the different parameters that we will need to solve the problem: characteristics of the studied line (number of buffers, minimum and maximum capacity of each buffer) and the parameters of the optimization algorithm. It is first of all necessary to enter the different data concerning the buffers, then the parameters of the SPEA-II algorithm (population size, archive size, number of generations, crossover and mutation probabilities).

Once all these data have been declared, a function is launched at the start of the simulation that initializes the parameters of the optimization algorithm (the archive in the case of the SPEA-II algorithm) and randomly generates the initial capacities of the buffers in the line. It also calls another function (*ModelLogic-RunBeginReplication*) that will be launched at the beginning of each replication. In this function, two cases arise; if the iteration number is less than the number of replications necessary to evaluate the performance (ten replications are necessary in our case, as mentioned earlier), the sub-function calling the optimization algorithm is not triggered. If, on the other hand, the number of necessary replications is attained, the data provided by the optimization algorithm are taken into account for a new evaluation, and so on until the stop criterion is met.

In our case, the total number of replications to be performed is calculated as follows: the number of replications for a performance evaluation (i.e. 10 replications) is multiplied by the size of the population in the SPEA-II genetic algorithm (i.e. 20 individuals) or the number of ants in the MOACS algorithm (i.e. 40 ants) and then multiplied by the number of generations of the algorithm (i.e. 100 generations). In our example, we therefore need 20,000 replications for the SPEA-II algorithm.

Changes to the buffer capacities and therefore the variable corresponding to the size of each buffer (e.g. Capbuffer 2 for buffer 2) are done in the VBA interface.

3.5.5.4. *Results and comparison*

In this section, we present the Pareto fronts obtained by the two algorithms after their application to the studied problem. Table 3.13 shows the Pareto fronts of both MOACS and SPEA2, as well as the mean and the standard deviation for each objective.

Algorithms	MOACS		SPEA2	
Objectives	Number of output products	Total capacity of the buffers	Number of output products	Total capacity of the buffers
Pareto front	276	78	276	75
	268	61	203	36
	172	22	185	29
	233	41	226	45
	257	54	264	61
	223	39	210	41
	244	46	240	49
	274	71	271	71
	252	50	250	53
	205	33	198	34
	213	36	-	-
	193	28	-	-
	184	26	-	-
Mean	230.30	45.00	232.30	49.40
Standard deviation	33.65	16.60	32.64	15.60

Table 3.13. *The numerical results for the studied problem*

	n_{f1}	n_{f2}	μ	μ^*	C_1	C_2
P1	13	10	-2.35	-0.68	10	58

Table 3.14. *A comparison of the MOACS and SPEA2 fronts*

Table 3.14 shows the comparison between the fronts given by each algorithm: $F1$ corresponds to the front given by the MOACS algorithm and $F2$ to that given by SPEA2.

The three comparison criteria are the number of solutions in each front, n_{fi}, the Riise distances μ and μ^* and the Zitzler distances C_i.

From these initial results, we may note that the MOACS algorithm gives a better performance than SPEA2. In fact, MOACS gives 13 solutions in the Pareto front, whereas SPEA2 gives 10 solutions.

Furthermore, the negative value of the Riise distance (μ^*) confirms that the Pareto front of MOACS is lower than that of SPEA2. We, therefore, have a greater number of output products with a smaller total buffer capacity. The third comparison criterion in turn proves the advantage of MOACS against SPEA2, as the Zitzler distance C_1 is smaller than C_2, meaning that we have more solutions from SPEA2 that are dominated by solutions from MOACS.

3.6. Layout

The problem of arranging or rearranging a workshop consists of determining the relative location of the different resources (machines, maintenance systems, handling systems, etc.) to facilitate the different flows that cross a workshop (the flow of raw materials, of semi-finished or finished products) in order to facilitate production while minimizing the related costs. One of the main costs that we consider is that of handling, with a view to improving the responsiveness of the workshop, to reduce delays, etc.

This section is inspired by several papers, as well as research projects – in particular those of N. Yalaoui [YAL 10] and Hani [HAN 06a].

The facility layout problem was presented by De Alvarenga *et al.* [ALV 00] as the search for the best arrangement of machines, equipment or departments in a space to optimize a temporal or spatial criterion (to save time and/or space) as well as to simplify the production system, thus improving the performance and productivity of an industry. A well thought out and optimized layout guarantees an efficient, responsive and productive system.

According to several authors, such as Proth in 1992 [PRO 92], there are several scenarios where the arrangement of a facility is to be considered. The most frequent are:

– a new organization in the facility;

– the construction of a new production structure;

– the optimization of the equipment allocation in the facility; and

– the addition of a workshop to a production unit.

3.6.1. *Types of facility layout*

The typology of layout systems uses several concepts, especially that of hierarchical division. Hierarchical decomposition consists of two large families: logical and physical.

3.6.1.1. *Logical layout*

This type of model involves the identification of flows and technologies and the ability to manage them. The procedure consists of grouping together the machines into cells so as to find the products manufactured according to the same schemes in the same cell. We then consider intracellular and extracellular flows. The effect of this decomposition is the maximization of the flows of materials and products in the cells as well as the minimization of the flow between cells. The interest of this type of structure is also the generation of technological blocks or delimited and manageable occupations.

3.6.1.2. *Physical layout*

The physical layout may be intracellular or extracellular. For an intracellular layout, the principle consists of finding the best arrangement of the machines in a previously established cell in a given area of the facility. There are several intercellular arrangements or layouts (Figure 3.19): circular, linear, multi-linear or nondescript [SOU 94].

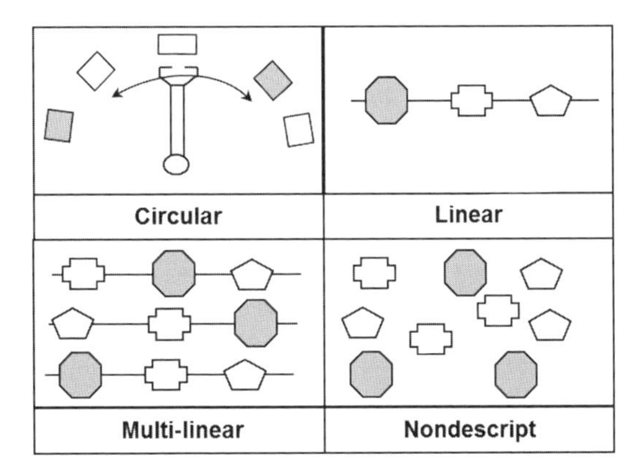

Figure 3.19. *Some types of cellular arrangement [SOU 94]*

Regarding the intercellular physical layout, the aim is to find the best arrangement of cells, departments or workstations inside the facility in order to minimize the costs of the layout, taking into account the geometries of the cells as well as that of the reception area of the facility. Mahdi [MAH 00] proposes the classification (Figure 3.20) of facility layout problems by breaking down the different facility types and characteristics.

3.6.2. *Approach for treating a layout problem*

Generally, constraints related to the geometries of the posts, cells or machines to be placed in the reception area of the facility or related to the system of processing products, as well as requirements of quality and industrial security, are clearly established. For the implementation of a rational occupation of space in a workshop, we should take into account [YAL 10, HAN 06a]:

– the use of both horizontal and vertical spaces in order to minimize the space used;

– the distances traveled by workers;

– the distances traveled by the products being manufactured, as well as the raw materials;

– the flexibility of the manufacturing processes, allowing adaptation to future changes in demand and to the diversification of the products; and

– the satisfaction of conditions of safety and production quality.

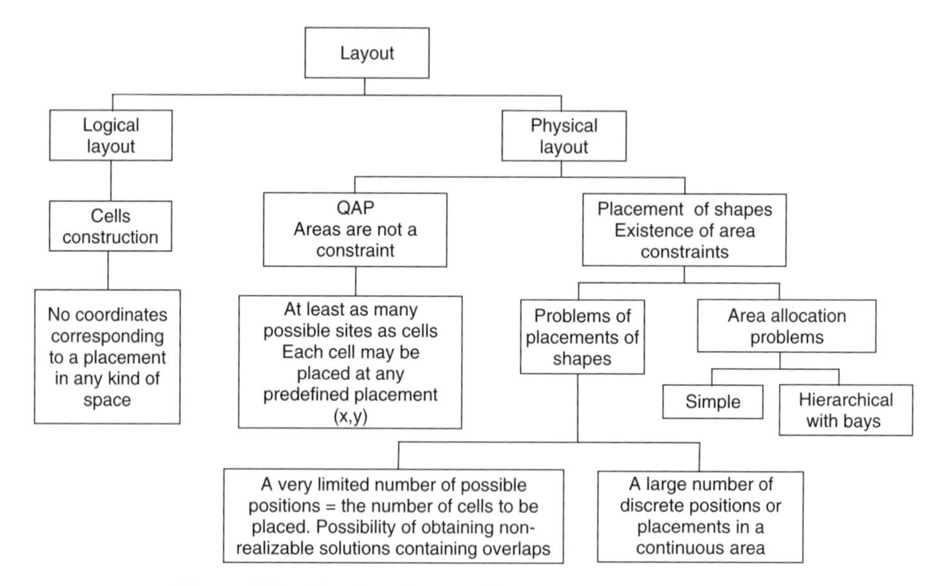

Figure 3.20. *The classification of layout problems [MAH 00]*

One may arrange a facility in several ways [HAN 06a]: linear, functional, cellular or fixed. In practice, one may find hybrid layouts – combinations of two or more types of layout.

3.6.2.1. *Linear layout*

The placement of workstations, departments or cells in the workshop in the order of the succession of production tasks is known as a linear layout. This arrangement is adopted when production occurs in series, where the range of products manufactured on a large scale is reduced, which is the case in flow shop-type facilities. This type of workshop is managed by the workflow. This is the case, for example in assembly lines etc. The complexity and diversity of production ranges may prevent the implementation of a single line, and thus the positioning of machines according to the order of the dominant range, so as to minimize the amount of backtracking necessary for the products.

3.6.2.2. *Functional layout*

An arrangement based on the grouping of equipment and installations according to their functionality is known as a functional layout. It is generally adopted for the manufacture of a large number of varied products in small quantities.

3.6.2.3. *Cellular layout*

The grouping of equipment in this type of arrangement is done according to the technological demands of the products. The objective is to group the resources of the workshop into cells. Each cell is dedicated to a group of products that primarily uses the resources of that cell. An application of this approach was implemented in 2010 by Yalaoui *et al.* [YAL 08, YAL 10]. The main objective of the concept of technological grouping [AUL 72] is to obtain families of products and allocate them to cells of machines. For this type of problem, a binary machine-products matrix is used as input data for the identification of the products processed by the machines.

3.6.2.4. *Fixed layout*

Several authors [HAN 06a, MAH 00] propose to apply this type of layout in the case of production in small quantities or when the product itself is fixed. A construction site where we arrange the construction equipment around a building is an example of this kind.

3.6.3. *The best-known methods*

Facility layout problems have been extensively addressed in the scientific literature, and there are commercial solutions in the market. In the hierarchy of decisions (strategic, tactical and operational), the problem of layout is said to be tactical, but it is clear that depending on the typology of the structure it may become operational, particularly in flexible configurations. This classification refers to the specifications that all methods must recognize the need to provide a solution. In sum, layout is considered as a frozen or fixed placement of equipment in a reception area. It has the goal of minimizing the flows of materials and products being manufactured in the facility. Problems of cellular layout (as defined earlier) are some of the most well-studied problems in the literature. A dynamic or flexible layout is an adaptive production system in constant evolution.

Facility layout was formulated mathematically for the first time by Koopmans and Beckman [KOO 57] in 1957. This formulation was based on a Quadratic Assignment Problem (QAP). This model is widely used to represent layout problems. QAP imposes fixed sites for the departments or cells, to be placed with product flows passing through them. QAP is used when the departments are of identical sizes or the possible sites are predetermined.

There are a series of studies and specific methods for solving numerous problems; we refer the reader to the literature. In the following, we give an example.

3.6.4. *Example of arranging a maintenance facility*

This example concerns the layout of workshops in a railway rolling stock maintenance facility [HAN 06a, HAN 07]. The facility has the role of ensuring the maintenance of passenger cars, which leads to the related activities of the repair, processing and manufacture of various parts. It had an annal production of a few hundred cars revised or processed and the reparation of tens of thousands of electric or electronic parts. With a workforce of several hundred workers carrying out some 50 different jobs (carpenters, boilermakers, mechanics, electricians, electronic engineers, etc.). The areas of operation range from the revision and modernization of cars to the repair of the components of these cars.

The company uses several production infrastructures (buildings), notably a main workshop for assembling and disassembling cars, as well as a set of workshops for the renovation of parts, storage facilities and facilities receiving raw materials and replacement parts. To perform these tasks, the company uses service-based organization: production and supply services. These two services work closely together in order to plan the needs of production.

The maintenance process is rather complex, and for reasons of confidentiality, is not fully detailed here in this work. It consists of a successive set of tasks necessary to carry out the maintenance of the cars. These tasks are listed below:

– arrival of cars into the establishment;

– disinfection and ventilation;

– appraisal;

– technical phase 1;

– disassembly;

– technical phase 2;

– reassembly;

– technical phase 3;

– technical phase 4;

– technical phase 5;

– reassembly of windows;

– brakes and electronics testing;

– weighing and cleaning;

– reassembly of accessories; and

– discharge.

The goal of studying the layout is to improve productivity and increase the production capacity of the establishment. The latter is composed of several buildings constructed on parallel tracks. The vehicles (train cars) to be processed arrive in lots and travel through the different buildings according to their previously defined work schedule. They are transported either by a carriage that crosses the tracks in the transverse direction on a fixed trajectory, or by a locomotive, in order to access the buildings that are not accessible by the transporter carriage.

Some work requires a significant processing time, which mobilizes sites. These tasks represent bottlenecks for the entire organization. The path of a car essentially passes by two buildings, written x_1 and x_2. Upon entry, the cars are treated on exterior tracks, then according to the associated work schedule, travel between both buildings. Finally, they finish their route on exit tracks. To model the problem, the authors divided each track into zones, each called a car site, where the maintenance tasks are carried out. According to the tasks, the car sites are distinguished into three possible types:

– if the site is crowded with product stocks, workbenches or offices, the site is said to be unusable;

– if the site is occupied with fixed resources (heavy machines), the site is said to be specialized, as the resource allocated to this site cannot be displaced; and

– if the site is not in either of the two previous scenarios, the site is said to be a general-purpose site.

The problem was modeled as in quadratic assignment. To optimize the production flow in the facility, we define a quadratic function Z (equation [3.89]) to be minimized.

The corresponding mathematical model is as follows:

$$Z : Minimize \sum_i \sum_j \sum_{i'} \sum_{j'} \sum_k \sum_w f_{r_{ij},r_{i'j'}}.D_{k,w}.P_{r_{ij},k}.P_{r_{i'j'},w} \quad [3.89]$$

under the constraints:

$$\sum_i \sum_j P_{r_{ij},k} = 1;\ \forall k \quad [3.90]$$

$$\sum_k P_{r_{ij},k} = 1;\ \forall i,\ \forall j \quad [3.91]$$

$$TE_k + TES_k = 1;\ \forall j = 1, \ldots, N-1 \quad [3.92]$$

– NE: the total number of sites in the facility;

– r_{ij}: resource j of task i, a task may be allocated to several resources;

– $D_{k,w}$: the distance between sites k and w; this is the number of uncrowded sites separating the two resources;

– $f_{r_{ij},r_{i'j'}}$: the production flow between resource r_{ij} and resource $r_{i'j'}$. The flow is calculated with respect to a given time frame, and expresses the quantity or volume of cars passing from one resource to another;

– $P_{r_{ij},k}$: a binary variable equal to 1 if r_{ij} is assigned to site k and 0 otherwise;

– TE_k: a binary variable equal to 1 if site k is general-purpose site and 0 otherwise; and

– TES_k: a binary variable equal to 1 if site k is specialized site and 0 otherwise.

Constraint [3.90] states the fact that, for a given site, a single resource is allocated to it. Constraint [3.91] expresses the fact that each resource is allocated to a single site and all of the resources must be allocated. Constraint [3.92] states that occupied sites are either general-purpose or specialized sites.

Contrary to other problems discussed in the literature, the flow matrix in this problem is not symmetric, i.e. $f_{ij} \neq f_{ji}$, and it is of size $(n+m)^2$ (n being the number of general-purpose sites and m the number of specialized sites). On the other hand, the distance matrix is symmetric, and of size n^2. All exchanges are done at the edges of the workshop corresponding to traveled distances equal to zero; the calculation of distance is done according to the minimum number of cars to be moved inside the building. Consequently, the distances for the workshop will have the values 0, 1 or 2.

Hybrid genetic algorithms are combined into a guided local search (GLS) [HAN 06b]. The steps of the overall method are:

– The encoding of the chromosome: since the workshop includes unusable spaces, this must be taken into consideration in the chromosome and represented by zero. The accepted sites are numbered according to their type of activity.

– Crossover: the crossover is used in the following way: two points in the chromosome are randomly selected and a child inherits elements that are found between the two crossover points of a parent. The remaining elements are from the other parent in the order in which they appear, starting from the second crossover point and avoiding the elements that already exist in the child.

– Mutation: the operation of mutation modifies the child, providing perturbations in order to vary the population. This operation is performed in this report with a probability of 0.1, which means that only 10% of the children are mutants.

– The application of the GLS algorithm to the QAP can be summarized as follows: taking an existing solution, a local search algorithm is first used (for example 2-opt). This first local search finds a local minimum, taking into account the increased function. If the minimum cost (non-increase) is less than the smallest cost, it is kept as the best found solution. Finally, the feature (the allocation) with the maximum use will have a penalty which increases.

The *GLS* method requires the calculation of a utility function $util$ (equation [3.93]) and an increased function $h(s)$ (equation [3.94]), where $g(s)$ is the cost function and λ is the parameter which allows the diversity of the search to be altered.

$$util(s, f_i) = I_i(s).\frac{c_i(s)}{1 + p_i} \quad [3.93]$$

$$h(s) = g(s) + \lambda.\sum_{i=1}^{n} I_i(s).p_i \quad [3.94]$$

Algorithm 3.3 The structure of the GLS–QAP algorithm [HAN 06b]

Step 1: Calculate s'
Step 2: Take the best solution s' as the initial solution s
Step 3: Apply the local 2-opt search taking into account the cost of the increased function; $s*$ will be considered as the solution with the smallest increased cost
if cost($s*$) < cost(s') **then**
  Replace s' with $s*$
  Find the characteristic (the allocation) of $s*$ with the maximum utility, write it as f_{i,π_i}
  Increase the corresponding penalty $p_{i,\pi_i} = p_{i,\pi_i} + 1$
end if
Step 4: Return to Step 3 until the stop criterion is met
s' is the best solution to the original problem

The application of the GLS algorithm to the QAP (Algorithm 3.3) is carried out as follows: the characteristic f_{i,Π_i} of solution s corresponds to the allocation of task i to site Π_i. The cost associated with the characteristic f_{i,Π_i} depends on the interaction of task i with all of the other tasks in solution s. This cost is calculated as shown in equation [3.95]:

$$C(i, \Pi_i) = \sum_{j=1}^{n} f_{ij}.D_{\Pi_i \Pi_j} \quad [3.95]$$

The value of λ best-adapted to the QAP is given by equation [3.96]:

$$\lambda = \frac{\sum_{i=1}^{n}\sum_{j=1}^{n} f_{ij}.\sum_{i=1}^{n}\sum_{j=1}^{n} D_{ij}}{n^4} \quad [3.96]$$

The algorithm was applied to a building containing 72 sites, of which 27 were unusable, 39 were specialized and 6 were general-purpose sites. There were 15 tasks to allocate and 27 resources in total. Initially, we assigned resources to sites in the order they exist in the workshop. The flow matrix describes the exchanges or the displacement of the production of one resource to another according to the work schedule of each type of product. This flow is calculated with respect to a given time frame. In our case, it is calculated for a year. The performance was improved by 19.76% with respect to the current implementation as it existed in the facility.

3.6.5. *Example of laying out an automotive workshop*

This study focuses on the business of designing, manufacturing and marketing spring clamps. An analysis of the company's production processes allows the identification of the different flows applied and the equipment present in the workshops. This leads to a tactical problem, focusing more specifically on the problem of the layout of the company's workshops. In order to detect and analyze all the problems related to the production flow, a map of the flows is drawn.

The reorganization of workshops follows a known process demonstrated by various studies available in the literature. This process consists of defining a workspace represented by the workshop, and potential sites for the machines. The general principle consists of assigning the different available resources to the different allocated spaces. A decrease in the distances between resources was considered as the objective. Several constraints may be taken into account, such as forbidden sites, the specificity of resources, etc.

The most widely used criterion in the literature is the minimization of the flow between cells. This requires the minimization of exceptions and empty spaces in the matrix. The exceptions (those outside of the cells, i.e. the necessity of using resources external to the cells) define the intercellular flow, and the empty spaces (the zeros in the cells) are associated with the density of the cells.

To solve the company's problem, a new approach was adopted. This is divided into three steps, and is called T-FLP. The first step consists of creating groups of machines and families of products based on the volumetric production flow. As a result, machine-product cells are obtained. The second step consists of allocating the machines to different positions according to their membership to the cells. An evaluation of the aggregate function combining the evaluation of the grouping and the layout is performed in the third step. These three steps are repeated several times, looping over the maximum number of possible cells. This approach represents the best compromise between a technological grouping maximizing the intracellular volumes and a physical layout minimizing the distance traveled between the machines. The steps of the solution are defined in Algorithm 3.4.

Algorithm 3.4 The solution method

1: **while** the maximum number of cells has not been reached **do**
2: STEP 1: Solve the problem of technological grouping
3: STEP 2: Solve the problem of layout taking into account the technological grouping solution
4: STEP 3: Evaluate the overall solution
5: **end while**

The industrial system is composed of 36 machines and 47 products. The volumes used are those of two years of production in terms of the number of pallets. The number of sites is equal to the number of machines.

In view of the obtained results, the loop on the cells varies between 3 and 36. The first step creates the cells using the formulation of the technological grouping equation. At the second step, the machines are allocated to the 36 positions selected using the FLP equation. Finally, at step 3, the solutions are evaluated using global assessment.

The solution obtained for technological grouping gives three cells with an intercellular volume of 100%. This represents 28,248 pallets and no transfer between cells. This solution is a reduction from the 17,349 pallets transferred between cells in the existing solution in the business. Compared to the existing solution, the algorithm developed by Yalaoui *et al.* [YAL 10] shows an improvement with a mean deviation of 34.5% on the total distance traveled by the forklifts.

Chapter 4

Tactical Optimization

4.1. Introduction

This chapter presents the aspects of the management of an industrial system in the medium term. It concerns decisions that are implemented on a time frame ranging from several weeks to several months. Depending on the industry in question and the characteristics of the demand (seasonal, growing, stable, etc.), the management of the industrial system must be adapted. Few systems work only on the basis of firm orders. The majority of medium-term management decisions to be made rely on demand forecasting. Demand forecasts may be refined over time so as to adapt to trends. They are now a part of the input data for the implementation of stock management policies, production and replenishment planning methods, as well as the management of quality control.

We first describe the main methods of demand forecasting, then the different stock management policies. Following this, we present the lot-sizing problems that are encountered during the management of production and replenishment planning. Finally, we discuss the principles of implementing quality control at reception as well as during manufacture.

4.2. Demand forecasting

4.2.1. *Introduction*

Demand forecasting aims to:

– estimate the probability of occurrence of future events;

– determine the temporal arrangement of these events;

– identify regularities in the series observed in the past;

– highlight relationship between the measured values.

Forecasting may be used in various domains such as marketing, production, replenishment, administration and control, as well as research and development. Numerous factors may influence the demand for a product, such as the general conditions of the market and the economic situation, the actions and reactions of competitors, as well as legislative actions (normalizations), the lifecycle of the product, technological innovations and changes in tastes and customs.

DEFINITION 4.1.– *Forecasting is the consideration of a key element of any management process: anticipation (forecasting).*

4.2.2. *Categories and methods*

Forecasting techniques may be qualitative or quantitative, and endogenous or exogenous.

Quantitative methods are based on the study of numerical data that characterize an economic variable. They let us highlight regularities that are likely to recur, and from which we may conceivably make forecasts about the future state of this variable. We may, for example, make forecasts about a shop's future sales by analyzing its past sales.

Qualitative methods are based on two aspects: the study of numerical data and in part, intuition. They let us highlight regularities in the past and take into account other aspects such as imagination, creativity and abilities in sociological, political or technical areas, which play an important role.

With the exception of (operational level) administrative problems, where they are sufficient, quantitative methods do not provide sufficient indications and reference points for other levels of management. Then, on a tactical level, and even more so on strategical levels, qualitative methods supplement quantitative methods.

Endogenous methods are based on the consideration of time as an explanatory variable, as well as the consideration of parameters within the system in question.

Each economic phenomenon contains an inherent inertial force, which leads it, independent of other economic variables (environment, etc.), to continue on its trajectory (future). The main modeling component of any system is the time factor, but note that time does not explain everything.

Exogenous methods are based on the consideration of time as an explanatory variable, like endogenous methods, as well as on other variables outside the system: microeconomic variables (advertising, price, characteristics and features of the product, etc.) or macroeconomic variables (policy, household incomes, domestic consumption, industrial production, etc.).

The choice of forecasting method depends mainly on the forecast time frame. If we are considering the short term, the methods are quantitative and endogenous, and in the long/medium term, the consideration of the qualitative aspect is essential.

4.2.3. *Time series*

The field of forecasting is inseparable from the notion of a time series.

DEFINITION 4.2.– *A time series is a sequence of observations of a certain phenomenon over time.*

The state of the phenomenon is observed at regular intervals and the periodicity depends on the possibilities of measurement and the practical (operational) conditions (considerations).

Time-series models allow us to identify the components and the composition rules.

The components are the following:

– the trend (T_t);

– the conjuncture (C_t);

– the seasonality (S_t);

– the residual (R_t).

DEFINITION 4.3.– *The trend is the mean long-term evolution of the phenomenon under consideration.*

It may be:

– fast: phone calls;

– slow: the car market;

– quasi-stable: cinema tickets;

– decreasing: coal mining.

The trend guides the evolution of the market, and it is an essential strategic variable.

DEFINITION 4.4.– *The conjuncture is a medium-term influential factor. It manifests as cyclic movements alternating over several years of growth and recession phases. The periodicity of these cycles is not constant.*

Multiple levels of conjuncture must be considered. The national conjuncture is superimposed over the specific conjuncture of a market. This is a complex factor to be understood in an indirect manner.

DEFINITION 4.5.– *The seasonality represents the fluctuations of constant periodicity (a year at most).*

Seasonal (annual) fluctuations may be attributable to the climate (for products such as fuel, swimsuits and skis) or to social, economic or cultural factors.

DEFINITION 4.6.– *The residual factor represents all the identifiable (controllable and therefore eventually manageable) disturbances on the one hand, and all random perturbations on the other hand.*

4.2.4. *Models and series analysis*

The values, observed at a date t, of the variable under consideration, written Y_t, are a function F of the values taken by the components at the same date:

$$Y_t = F(T_t, C_t, S_t, R_t) \tag{4.1}$$

There are two main composition models: the additive model (AM) and the multiplicative model.

In the case of an additive model, we have the function F expressed in the following way:

$$Y_t = F(T_t, C_t, S_t, R_t) = T_t + C_t + S_t + R_t \tag{4.2}$$

Each element is measured in the same units as Y_t. The variation of one of the components is independent of the values taken by the others. If we consider m periods, we have:

$$\sum_{t=i+1}^{i+m} S_t = 0 \tag{4.3}$$

For a multiplicative model, Y_t is measured in the same units as the trend. The other components are integrated as factors:

$$Y_t = F(T_t, C_t, S_t, R_t) = T_t \times C_t \times S_t \times R_t \tag{4.4}$$

Over m periods, we then have:

$$\sum_{t=i+1}^{i+m} S_t = m \quad [4.5]$$

Note that a multiplicative model can be turned into an AM using a logarithmic transformation.

The analysis of a series consists of identifying its different components using a decomposition procedure. The most common technique consists of separating out the medium-term evolution of the phenomenon (the combination of the trend and the conjuncture) using the linear regression or the moving average method, or a hybrid method (a combination of various methods). We may then deduce the seasonal coefficients S and the residual component R.

EXAMPLE 4.1.– We will consider the evolution table (Table 4.1) of a variable Y_t over the last 12 months (periods).

Period t	1	2	3	...	12
Y_t	Y_1	Y_2	Y_3	...	Y_{12}

Table 4.1. *The evolution of the variable over 12 periods*

4.2.4.1. *Additive models*

If we wish to develop an AM, we must first, with a graphical method, identify the periodicity of the phenomenon. Suppose that we observe a period of $n = 4$ time units. As we are working over 12 units, we have three periods composed of four time units each.

Secondly, we must seasonally adjust the data using centered moving averages (CMAs).

For $j \geq (\frac{n}{2} + 1)$ and n even:

$$\text{CMA}(j) = \frac{(\frac{1}{2}Y_{j-n/2} + \ldots + Y_{j-1} + Y_j + Y_{j+1} + \ldots + \frac{1}{2}Y_{j+n/2})}{n} \quad [4.6]$$

For $j \geq (\frac{n+1}{2})$ and n odd:

$$\text{CMA}(j) = \frac{(\frac{1}{2}Y_{j-\frac{n-1}{2}} + \ldots + Y_{j-1} + Y_j + Y_{j+1} + \ldots + \frac{1}{2}Y_{j+\frac{n-1}{2}})}{n} \quad [4.7]$$

In the previous example, n is even, and the CMA is calculated from $j = 3$ up to $j = 10$. We have, for example:

$$\text{CMA}(3) = \frac{\frac{1}{2}Y_1 + Y_2 + Y_3 + Y_4 + \frac{1}{2}Y_5}{4} \quad [4.8]$$

$$\text{CMA}(4) = \frac{\frac{1}{2}Y_2 + Y_3 + Y_5 + Y_5 + \frac{1}{2}Y_6}{4} \quad [4.9]$$

$$\text{CMA}(10) = \frac{\frac{1}{2}Y_8 + Y_9 + Y_{10} + Y_{11} + \frac{1}{2}Y_{12}}{4} \quad [4.10]$$

Following this, we determine the trend of the CMAs using simple linear regression. Thus, we calculate the parameters of a line:

$$T(t) = \hat{a}t + \hat{b} \quad [4.11]$$

$$\hat{a} = \frac{\text{COV}(t, \text{CMA})}{V(t)} = \rho\sqrt{\frac{V(\text{CMA})}{V(t)}} \quad [4.12]$$

$$\hat{b} = Moy(\text{CMA}) - \hat{a}Moy(t) \quad [4.13]$$

We may then calculate the seasonal coefficients using the following formula:

$$S(t) = Y(t) - T(t) \quad [4.14]$$

Next, we evaluate the uncorrected adjusted seasonal coefficients:

$$S^*(t+n) = S^*(t+2n) = S^*(t+Kn) \;\; \forall K \in \mathbb{N} \quad [4.15]$$

In our example, this implies that:

$$S^*(1) = S^*(5) = S^*(9) \quad [4.16]$$

$$S^*(2) = S^*(6) = S^*(10) \quad [4.17]$$

$$S^*(3) = S^*(7) = S^*(11) \quad [4.18]$$

$$S^*(t+jn) = \frac{\sum_{j=0}^{K-1} S^*(t+jn)}{K} \;\; \forall j < K \text{ and } t < n \quad [4.19]$$

The following step consists of calculating the correction parameter A:

$$A = \frac{\sum_{j=1}^{n} S^*(j)}{n} \quad [4.20]$$

This lets us calculate the corrected seasonal coefficients:

$$\overline{S(t)} = S^*(t) - A \;\; \forall t \quad [4.21]$$

The forecast is then calculated with the following formula:

$$P(t) = \overline{S}(t) + T(t) \;\; \forall t \in [1; +\infty[\qquad [4.22]$$

The estimation of the errors is done using past data about the K seasons with period n:

$$R(t) = |Y(t) - P(t)| \qquad [4.23]$$

EXAMPLE 4.2.– Let us consider an AM. The first two columns of Table 4.2 indicate the periods and the observed data (Y). From these data, we may calculate the CMA column. The trend (T) is then evaluated, and then the seasonal coefficients ($S(t)$) and the corrected coefficients ($S^*(t)$) are evaluated. Note that the correction parameter A in this example is 106.66. We may then obtain forecasts and compare them with the actual outcomes. The mean error over the 12 periods is 78.58.

Period t	$Y(t)$	CMA	T	S	$S^*(t)$	$\overline{S}(t)$	$P(t)$	Error
1	3,150		3,149.55	0.45	−38.48	−45.14	3,104.41	45.59
2	3,200		3,243.25	−43.25	38.33	31.66	3,274.91	74.91
3	3,399	3,275.6	3,336.94	62.06	208.88	202.21	3,539.16	140.16
4	3,162	3,386	3,430.64	−268.64	−182.06	−188.73	3,241.91	79.91
5	3,533	3,533.75	3,524.34	8.66	−38.48	−45.14	3,479.19	53.81
6	3,700	3,686.13	3,618.08	81.97	38.33	31.66	3,649.69	50.31
7	4,081	3,789.13	3,711.73	369.27	208.88	202.21	3,913.94	167.06
8	3,704	3,862.13	3,805.42	−101.42	−182.06	−188.73	3,616.69	87.31
9	3,810	3,911.88	3,899.12	−89.12	−38.48	−45.14	3,853.97	43.97
10	4,007	3,950	3,992.81	14.19	38.33	31.66	4,024.47	17.47
11	4,172	4,025.5	4,086.51	85.49	208.88	202.21	4,288.72	116.72
12	3,918	4,131.88	4,180.21	−262.21	−182.06	−188.73	3,991.47	73.47
13	4,200	4,265.5	4,273.90	−73.90	−38.48	−45.14	4,228.76	28.76
14	4,468	4,409.13	4,367.60	100.40	38.33	31.66	4,399.26	68.74
15	4,780		4,461.29	318.71	208.88	202.21	4,663.51	116.49
16	4,459		4,554.99	−95.99	−182.06	−188.73	4,366.26	92.74

Table 4.2. *An example of forecasts over 12 periods with an additive model*

4.2.4.2. *Multiplicative model*

The first step is identical to the case of the AM, concerning the identification of the periodicity of the phenomenon. Next, we seasonally adjust the data with CMAs, as before. Following this, we determine the trend of the CMAs using simple linear regression. We thus calculate the parameters of a line:

$$T(t) = \hat{a}t + \hat{b} \qquad [4.24]$$

where

$$\hat{a} = \frac{\text{COV}(t, \text{CMA})}{V(t)} = \rho\sqrt{\frac{V(\text{CMA})}{V(t)}} \quad [4.25]$$

where $V(t)$ is the variance.

Following this, we calculate the seasonal coefficients using the following formula:

$$S(t) = \frac{Y(t)}{T(t)} \quad [4.26]$$

The uncorrected adjusted seasonal coefficients are calculated with:

$$S^*(t) = S^*(t+n) = S^*(t+2n) = \ldots = S^*(t+Kn) \quad [4.27]$$

Considering the same example as for the AM (Table 4.1), we obtain $S^*(1) = S^*(5) = S^*(9)$, $S^*(2) = S^*(6) = S^*(10)$ and $S^*(3) = S^*(7) = S^*(11)$. $S^*(13)$, $S^*(14)$ and $S^*(15)$ do not exist. Next, we calculate the correction parameter A:

$$A = \frac{n}{\sum_{j=1}^{n} S^*(j)} \quad [4.28]$$

The adjusted seasonal coefficients are calculated with:

$$\overline{S}(t) = S^*(t) \times A \;\; \forall t \quad [4.29]$$

For the previous periods as well as for the future periods, the forecasts are obtained by the following formula:

$$P(t) = \overline{S}(t) \times T(t) \;\; \forall t \in [1, +\infty[\quad [4.30]$$

Finally, we may estimate the errors of the model by calculating $R(t) = |Y(t) - P(t)|$.

EXAMPLE 4.3.– We consider the same numerical example as for the AM, and develop the associated multiplicative model (Table 4.3). The correction parameter A in this example equals 16.02. Comparison of the forecasts with the actual outcomes shows a mean error of 77.09 over the 12 periods. Therefore, we conclude that the multiplicative model is more efficient than the additive model, with a lower mean error.

4.2.4.3. *Exponential smoothing*

The method of exponential smoothing is used to estimate the future evolution of phenomena considered on a very short period of time (at most 6 months). To do this, we generally have three methods: moving averages, the Box–Jenkins method and the (simple or double) exponential smoothing method.

Period t	$Y(t)$	CMA	T	$S(t)$	$S^*(t)$	$\overline{S}(t)$	$P(t)$	Error
1	3,150		3,149.55	1.00	0.99	0.98	3,116.09	33.90
2	3,200		3,243.25	0.98	1.00	1.00	3,268.23	68.23
3	3,399	3,275.62	3,336.94	1.019	1.05	1.05	3,508.11	109.11
4	3,162	3,386	3,430.63	0.92	0.95	0.95	3,264.66	102.66
5	3,533	3,533.75	3,524.33	1.00	0.99	0.98	3,486.89	46.10
6	3,700	3,686.75	3,618.03	1.02	1.00	1.00	3,645.91	54.08
7	4,081	3,789.12	3,711.72	1.09	1.05	1.05	3,902.12	178.87
8	3,704	3,862.125	3,805.42	0.97	0.95	0.95	3,621.32	82.67
9	3,810	3,911.875	3,899.11	0.97	0.99	0.98	3,857.69	47.69
10	4,007	3,950	3,992.81	1.00	1.00	1.00	4,023.58	16.58
11	4,172	4,025.5	4,086.51	1.02	1.05	1.05	4,296.13	124.13
12	3,918	4,131.87	4,180.20	0.93	0.95	0.95	3,977.97	59.97
13	4,200	4,265.5	4,273.90	0.98	0.99	0.98	4,228.49	28.49
14	4,468	4,409.12	4,367.59	1.02	1.00	1.00	4,401.125	66.74
15	4,780		4,461.29	1.07	1.05	1.05	4,690.14	89.85
16	4,459		4,554.98	0.97	0.95	0.95	4,334.62	124.37

Table 4.3. *An example of forecasts over 12 periods with a multiplicative model*

Exponential smoothing was introduced by R.G. Brown in 1959. The forecasts obtained by this method are based on the evaluation of the mean of the previous results. Weighted averages are established, where the weight of an observation decreases with its age. In simple exponential smoothing (SES), we distinguish between the cases where:

– the trend is zero and there is no seasonality;

– the trend is zero and there is some seasonality.

In DES, we distinguish between the cases where:

– the trend is non-zero and there is no seasonality;

– the trend is non-zero and there is some seasonality.

4.2.4.3.1. Simple exponential smoothing

SES consists of sorting the trend from the fluctuations. If the result observed over a period is greater than the expected average, this may be caused by:

– the progression of the trend;

– variations due to uncontrollable noise.

The procedure to be adopted is:

– For the former, we predict a higher mean (forecast) over the following period.

– For the latter, we do not change the mean (forecast) as the phenomenon is not controllable.

We write the forecast evolution of a period t as $\hat{Y}_t$. Y_t is the value that is actually observed. The discrepancy between the forecast and the observed value for a period t is written as $\Delta = Y_t - \hat{Y}_t$.

Thus, exponential smoothing, in general, takes into account the discrepancy over the period t for the forecasts over the period $(t+1)$:

$$\hat{Y}_{t+1} = \hat{Y}_t + \alpha \times \Delta = \hat{Y}_t + \alpha \times (Y_t - \hat{Y}_t) \quad [4.31]$$

It should be noted that α is a parameter that determines the sensitivity of the forecast. If α is low, the model is not very sensitive to the fluctuations (or completely insensitive if $\alpha = 0$). If α is high, the model fluctuates greatly, rapidly integrating the observed deviations. Note that if $\alpha = 1$, the next period will be identical to the current period.

It should also be noted that when $\Delta \geq 0$, the variation increases, and when $\Delta \leq 0$, the variation decreases.

From equation [4.31], we may write that:

$$\hat{Y}_{t+1} = \alpha Y_t + (1-\alpha)\hat{Y}_t \quad [4.32]$$

Therefore, we have the recursive relation that, for all values of t, is:

$$\hat{Y}_t = \alpha Y_{t-1} + (1-\alpha)\hat{Y}_{t-1} \quad [4.33]$$

From the above equation, we have:

$$\hat{Y}_{t+1} = \alpha Y_t + \alpha(1-\alpha)\hat{Y}_{t-1} + (1-\alpha)^2\hat{Y}_{t-1} \quad [4.34]$$

in general terms; we then have:

$$\hat{Y}_{t+1} = (1-\alpha)^{n+1}\hat{Y}_{t-n} + \sum_{i=0}^{n}(\alpha(1-\alpha)^i\hat{Y}_{t-i}) \quad [4.35]$$

when $n \to +\infty$; we then have $(1 - \alpha \to 0)$; and we hence obtain:

$$\hat{Y}_{t+1} = \sum_{i=0}^{\infty}(\alpha(1-\alpha)^i\hat{Y}_{t-i}) \quad [4.36]$$

Note that the exponential smoothing is much greater with a small α coefficient. For example, for a coefficient $\alpha = 0.1$, the previous observations of the 10 precedents contribute 35% of the forecast. For a coefficient $\alpha = 0.5$, the previous observations of the 10 precedents contribute 0.1% of the forecast. They are almost not taken into account.

EXAMPLE 4.4.– Let us take the example in Table 4.4. This concerns the establishment of forecasts over 12 periods with the SES model. The value of α is fixed at 0.2. In this example, the mean error between the forecasts and the outcomes is 333.32.

Period t	$Y(t)$	Forecast $Y(t)$	Error
1	3,150	3,150.00	0.00
2	3,200	3,150.00	50.00
3	3,399	3,160.00	239.00
4	3,162	3,207.80	−45.80
5	3,533	3,198.64	334.36
6	3,700	3,265.51	434.49
7	4,081	3,352.41	728.59
8	3,704	3,498.13	205.87
9	3,810	3,539.30	270.70
10	4,007	3,593.44	413.56
11	4,172	3,676.15	495.85
12	3,918	3,775.32	142.68
13	4,200	3,803.86	396.14
14	4,468	3,883.09	584.91
15	4,780	4,000.07	779.93
16	4,459	4,156.06	302.94

Table 4.4. *An example of forecasts over 12 periods with a simple exponential smoothing model*

4.2.4.3.2. Double exponential smoothing

DES consists of sorting the fluctuations from the trend when the latter is non-negligible. Indeed, if the trend is strong with SES, then there will be an increasing discrepancy between the forecasts and the outcomes. For the DES model, the trend of the phenomenon follows a linear variation:

$$T_t = a_t + b \qquad [4.37]$$

When $n \to \infty$, we have $(1-\alpha)^{n+1} \to 0$. In this case, we may express the forecast for period $t+1$ as:

$$\hat{Y}_{t+1} = \sum_{i=0}^{\infty}(\alpha(1-\alpha)^i[b + a(t-i)] \qquad [4.38]$$

$$\Leftrightarrow \hat{Y}_{t+1} = \sum_{i=0}^{\infty}(\alpha(1-\alpha)^i b) + \sum_{i=0}^{\infty}(\alpha(1-\alpha)^i at) - \sum_{i=0}^{\infty}(\alpha(1-\alpha)^i ai) \qquad [4.39]$$

Note that $\sum_{i=0}^{\infty}(\alpha(1-\alpha)^i) = 1$ and $\sum_{i=0}^{\infty}(\alpha(1-\alpha)^i i) = (1-\alpha)/\alpha$. Hence, we may deduce that:

$$\hat{Y}_{t+1} = b + at - a\frac{1-\alpha}{\alpha} \qquad [4.40]$$

We find that the discrepancy between the model's estimation for $t+1$ and the observed value for the period t equals $a\frac{1-\alpha}{\alpha}$. This leads to a total error between the forecast for $t+1$ and the outcome for $t+1$ calculated as follows:

It is known that:

$$T_{t+1} = Y_{t+1} = a(t+1) + b = at + b + a \quad [4.41]$$

Thus, the error between the DES forecast and the model is:

$$R = Y_{t+1} - \hat{Y}_{t+1} = (at + a + b) - (b + at - a\frac{1-\alpha}{\alpha}) \quad [4.42]$$

that is

$$R = a + a\frac{1-\alpha}{\alpha} = \frac{a}{\alpha} \quad [4.43]$$

If, with the same coefficient, we smooth the initial forecasts, we obtain a doubly smoothed series:

$$\overline{\overline{Y}}_{t+1} = \alpha\hat{Y}_{t+1} + (1-\alpha)\overline{\overline{Y}}_t \quad [4.44]$$

and:

$$\overline{\overline{Y}}_{t+1} = \sum_{i=0}^{\infty}(\alpha(1-\alpha)^i\hat{Y}_{t+1-i} \quad [4.45]$$

which may be written as:

$$\overline{\overline{Y}}_{t+1} = b + at - 2a\frac{1-\alpha}{\alpha} \quad [4.46]$$

We may note that there is the same discrepancy between Y_t and $\hat{Y}_{t+1}$ as between $\hat{Y}_{t+1}$ and $\overline{\overline{Y}}_{t+1}$.

To correct the forecast, we must then add the calculated difference between the smoothed value and the doubly smoothed value for the considered period, plus a, which is the increase in the trend from one period to another, to $\hat{Y}_{t+1}$. We then have:

$$\hat{Y}_{t+1} - \overline{\overline{Y}}_{t+1} = a\frac{1-\alpha}{\alpha} \quad [4.47]$$

then:

$$a = \frac{\alpha(\hat{Y}_{t+1} - \overline{\overline{Y}}_{t+1})}{1-\alpha} \quad [4.48]$$

The forecast for the period $t+1$ is then given by:

$$Y_{t+1}^p = 2\hat{Y}_{t+1} - \overline{\overline{Y}}_{t+1} + \frac{\alpha\lfloor\hat{Y}_{t+1} - \overline{\overline{Y}}_{t+1}\rfloor}{1-\alpha} \quad [4.49]$$

EXAMPLE 4.5.– Table 4.5 shows an example of the application of the DES model. It concerns forecasts made over 12 periods. The value of α is fixed at 0.2. In this example, the mean error between forecasts and outcomes equals 131.87. The DES model is thus more efficient than the SES model, with a less significant mean error.

Period t	$Y(t)$	Forecast $Y(t)$	Error	Double $Y(t)$	Forecast	Error
1	3,150	3,150.00	0.00	3,150.00	3,150.00	0.00
2	3,200	3,150.00	50.00	3,150.00	3,150.00	50.00
3	3,399	3,160.00	239.00	3,152.00	3,170.00	229.00
4	3,162	3,207.80	−45.80	3,163.16	3,263.60	−101.60
5	3,533	3,198.64	334.36	3,170.26	3,234.12	298.88
6	3,700	3,265.51	434.49	3,189.31	3,360.77	339.23
7	4,081	3,352.41	728.59	3,221.93	3,515.51	565.49
8	3,704	3,498.13	205.87	3,277.17	3,774.33	−70.33
9	3,810	3,539.30	270.70	3,329.59	3,801.44	8.56
10	4,007	3,593.44	413.56	3,382.36	3,857.29	149.71
11	4,172	3,676.15	495.85	3,441.12	3,969.94	202.06
12	3,918	3,775.32	142.68	3,507.96	4,109.52	−191.52
13	4,200	3,803.86	396.14	3,567.14	4,099.75	100.25
14	4,468	3,883.09	584.91	3,630.33	4,199.03	268.97
15	4,780	4,000.07	779.93	3,704.28	4,369.81	410.19
16	4,459	4,156.06	302.94	3,794.63	4,607.83	−148.83

Table 4.5. *An example of forecasts over 12 periods with a double exponential smoothing model*

4.3. Stock management

In the current context of the globalization of flows, good stock management takes into account the significant progress of production and manufacturing techniques. Stock management is relevant at different levels of the logistics chain, from the suppliers' suppliers to the customers' customers via other actors such as producers, carriers, wholesalers and retailers. This management occurs during the replenishment and production, as well as at the shipping and distribution stages.

The dilemma for every manager, whether they are a provider, producer or vendor, is not only to obtain a service rate of 100% to satisfy all their clients, but also to have the lowest possible storage cost. Indeed, the tendency is to hold a large amount of stock at a high storage cost.

The storage of products constitutes a break in the physical flow of the logistics chain. This break in the flow occurs at every echelon of the chain, from the moment the raw materials are received to the moment they are delivered to customers as finished products. These breaks are, of course, detrimental to a just-in-time policy, where non-productive time must be removed, but are a necessary evil as they limit breaks in replenishment or delivery. They thus serve as buffers between each step of production and distribution.

4.3.1. *The different types of stocked products*

The classification of products can be done according to the demand. Demand may be dependent, and deduced from another item, or independent. Dependent demand is used for finished products which depend on a set of semi-finished products, components and bills of material. Then, this classification is done according to the stage of completion of the product in the production process. At the beginning of this process, we find stocks of raw material. These are the set of received items that have not entered into the manufacturing process, such as materials, components or purchased subassemblies. These stocks allow us to:

– speculate and anticipate fluctuations in price;

– obtain a reduction in acquisition costs;

– be protected against the failures of suppliers and deliveries (time, quantity and quality);

– perform quality control.

Then, once the production process is initiated, it is called works in progress. This group contains the set of items incorporated into the manufacturing process, and undergoing transformation. These stocks give us:

– a partitioning of the manufacturing process into intermediate steps;

– protection against production downtime and failures.

Finally, once the production process is completed, the products are called finished products, which include the set of items ready to be dispatched. These stocks allow us to:

– reduce delays in delivery to the customer;

– cushion against fluctuations in demand (seasonal, random);

– achieve a better use of production tools;

– create safety stocks to deal with an unexpected customer demand.

We may also distinguish stocks according to the types of product, such as:

– Spare parts and miscellaneous provisions. This is the set of items used during production, but which do not make up a part of the products and the bill of materials.

– Waste, scrap and leftovers from manufacture.

These two types of stocks have long been neglected in the production process, but they are currently a point of interest in the optimization of logistics flow. Indeed, stocks of maintenance parts will directly influence the availability of production resources (machines or handling equipment), and stocks of waste or scrap may affect the production flow, either by physical hindrance or by the occupation of handling equipment.

4.3.2. *The different types of stocks*

Depending on the characteristics of the production system, we may have different types of stock. When we wish to anticipate future demand or when the manufacturing time is longer than the customer waiting time, a manager will stockpile in anticipation. Then, when we wish to deal with the variability of demand or delays in delivery, and thus avoid breaks in consumption, we will create fluctuation or protection stocks. We may also have stocks that are managed by batch size when there is a difference between the quantity desired and that ordered. The desired quantity most often corresponds to the economic quantity that will allow us to reduce shipping or production starting costs. When we wish to avoid breaks between the source and the delivery point, i.e. when the time to transport a product is significant, we talk of transport stocks. Finally, we speak of cover stocks for certain products that are subject to fluctuations due to variations related to financial markets, or climatic conditions. We may cite the examples of wood, cereals and certain metals (stainless steel, copper and aluminum). These stocks allow for speculation, and sale at the best price.

4.3.3. *Storage costs*

All management of a process in an industrial context involves the minimization of costs. Good stock management therefore includes a reduction in these costs while assuring an optimal service to the customer. To establish an appropriate stock policy, which can be adapted by the market, it is necessary to consider the various costs associated with the detention of products, the placement of orders for these products and their shortage. Of course, we must not forget the cost of purchasing the products, but this is more related to procurement than to storage.

Ownership or holding costs, written as C_d, are costs related to the storage place, and are generated by:

– the rent or depreciation of the warehouse (premises acquisition cost);

– the deprecation of the premises' equipment (handling apparatus);

– operational costs: lighting, heating, insurance and property taxes;

– personnel costs related to the stock (salaries and expenses);

– costs generated by obsolescence, damage and theft;

– stock management costs information technology (IT).

These are expressed as a percentage of the mean annual value of the stock, and may vary between 20% and 30% according to the stored products.

Acquisition costs, written as C_a, also called ordering costs, are the expenses related to the placing of the order. These represent:

– The costs generated by the establishment or renewal of the stock.

– The salaries and expenses of commercial service agents and purchases, such as market research, monitoring, recovery of deadlines, negotiations, modification of the goods, qualitative and quantitative controls and classification of invoices.

– The salaries and expenses of accountants such as records, the payment of invoices and the entry of the products into the stock.

– Operating expenses of these services (rent, lighting, supplies, postage, travel, IT (stock management)).

Acquisition costs are mostly constituted by salaries and expenses, and may be significant due to the openness of global markets. Indeed, this openness generates additional costs due to travel to foreign countries, the hiring of bilingual salesmen or interpreters, the use of legal advisors for the drafting and enforcement of clauses in contracts (advisors during trials), litigation (delays, quality, compliance) and the use of international transport: customs, international payments. The cost of placing an order can be estimated by the following formula:

$$\text{Cost of an order} = \frac{\text{Total acquisition costs}}{\text{Annual no. of commands}}$$

Next, we describe the costs of stockouts, written as C_r, which are generated by the absence of the product when it is ordered. These stockouts imply unrealized revenue or delay penalties. They may lead to a low customer service level and may affect the company in a rough competition context.

We consider two stockout cost scenarios: shortage of products ready for distribution or shortage of raw materials during production. In the first case, sales are lost if the customer goes to a competitor, or deferred if he or she agrees to wait. In the second case, deliveries may be lost, and it is necessary to launch a new order procedure. We may eventually call upon another provider or stop production if we cannot fill the gap quickly. In the case where delivery can be deferred, it is necessary to modify the production plan.

If it is quite easy to calculate holding and ordering costs, it is much more difficult to calculate stockout costs because they depend on the perturbation to the flow of production and distribution.

In storage costs, there are also shipping costs that come from the preparation of orders, packaging, loading, transport and the salaries of the shipping service. They depend on the logistic structure (warehouses, delivery routes, modes of transport, storage method, etc.).

4.3.4. *Stock management*

Stock management allows the accomplishment of two main tasks:

– The monitoring (or review) of stocks, which consists of the knowledge of the physical level and the amount of stocked items through the functions of warehousing and development.

– The optimization of stocks, which lets us increase the level of service and decrease costs through methods based on the Pareto approach, classification modes and also replenishment methods.

However, before addressing the different methods of stock management, it is necessary to specify the functioning of a stock.

4.3.4.1. *Functioning of a stock*

The general functioning of a stock starts at an initial level and decreases progressively due to its outputs. These outputs are called consumption or demands. After reaching a certain (positive, zero or negative) level, which is monitored during inspection periods, a replenishment order of a quantity Q is initiated. The arrival of the ordered products occurs after a delivery time. This management is characterized by a certain periodicity which is the time between two inputs. Figure 4.1 gives an example of the evolution of a stock. As soon as the stock reaches zero, an order of a fixed quantity Q is placed. Since consumption is constant, orders are made with a fixed period. The mean stock here is $Q/2$.

In the simplest case, consumption is constant, the stock inspection is continual and delivery times are zero, which generates fixed order quantities. This management may be more complex, however, when demand and delivery times are random and it is difficult to characterize their behavior with mathematical laws. Here are some definitions.

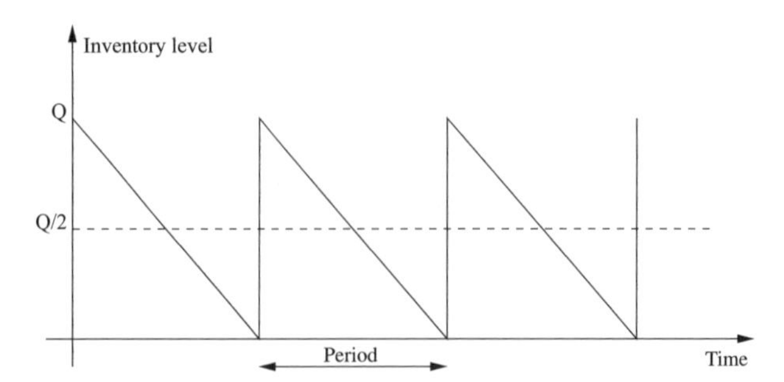

Figure 4.1. *The evolution of a stock level*

DEFINITION 4.7.– *We define a rolling stock (S_r) = initial stock (S_i) − stock at the end of a period(S_f).*

DEFINITION 4.8.– *The mean stock is calculated as follows: (S_m) = rolling stock (S_r) + safety stock (SS).*

To be able to determine the management method for a stock, we make use of several operational indicators.

DEFINITION 4.9.– *The service rate is a measure of the quality of the provisions from suppliers. It gives us an indicator of unfulfilled deliveries:*

$$SR = \frac{\textit{No. of orders received}}{\textit{No. of orders expected}}$$

We also refer to the calculation of the service rate when we wish to obtain a measure of the delivery efficiency with respect to customers, and to have an indicator of the number of stockouts. It is then calculated in the following way:

$$SR = \frac{\textit{Consumption in a period}}{\textit{Demand in the period}}$$

DEFINITION 4.10.– *The turnover of a stock is a measure of its efficiency, and is calculated as the ratio of the output amount to the mean value:*

$$TR = \frac{\textit{Annual consumption}}{\textit{Mean stock}}$$

DEFINITION 4.11.– *The coverage of a stock is:*

$$CV = \frac{\textit{No. of days in the year}}{\textit{Turnover}}$$

EXAMPLE 4.6.– Consider the following example, where we have a mean stock of 500 items, a yearly consuption of 3,000 items and a work period of 365 days. Thus, we obtain:

$$TR = \frac{3,000}{500} = 6$$

$$CV = \frac{365}{6} = 61\ days$$

Therefore, we have a turnover of 6, which means that the stock is totally replaced six times per year, and a coverage of 61 days, which means that, if demand is constant, we will have 61 days without stockouts.

DEFINITION 4.12.– *We may also define the turnover for a set of products. This is calculated as follows:*

$$TR = \frac{\textit{Sum of consumption}}{\textit{Sum of the mean values of stocks}}$$

$$TR = \frac{\sum_i D_i \times u_i}{\sum_i D_{mi} \times u_i} \quad [4.50]$$

where:

– D_i*: the annual consumption of item* i*;*

– D_{mi}*: the mean stock of* i*;*

– u_i*: the unit cost of* i*;*

– $TR_i = \frac{D_i}{D_{mi}}$ *and* $CV_i = \frac{365}{TR_i}$.

EXAMPLE 4.7.– Consider the data in Table 4.7. The turnover of the set of products is then $TR = 15.78$ and the coverage is $CV = 23$ days.

Item	D_i	D_{mi}	u_i	TR_i	CV_i
1	1,000	100	30	10	36
2	2,000	50	10	40	9
3	5,000	250	5	20	18

4.3.4.2. *Stock monitoring*

The monitoring of stocks is divided into two parts. The first part is warehousing, which consists of the placement of items in stores before their removal for shipment to the customer, and the other part involves stock valuation, which aims to calculate the different costs of the items.

If we break down the warehousing operations, the first action is the reception of products, which consists of verifying the quantity, the compliance and recording the input movement (on forms or using IT). Next comes arrangement, which is the physical placement of items, the item and its place being recorded (on forms or in a computer database). We then consider issue, which consists of physically withdrawing items for an order, and recording their exit (on form or in a computer). All of these operations are sources of errors, and it is necessary to regularly verify the existence or non-existence of the products in the stock. It is therefore necessary to add an inventory operation to each of these operations. An inventory consists of comparing the physical stocks with computerized stocks in order to recognize items and verify their quantities. Article L.123-12 of the commercial code imposes an annual physical inventory by and for traders. This may, in general, be done by counting at the end of the season, or in a continual manner by recording inputs and outputs if we have a good IT system, and much rigorously on the part of the managers. The inventory may also be done on a rotation basis. Each day, small quantities within the boundaries of a premise are counted so that there is no disturbance in the logistic flow.

4.3.4.3. *Stock valuation*

The valuation of stock movements aims to determine the value of the items in stock, i.e. their unit cost. To do this, it is necessary to take into account the movements that take place in the stock. The stock level is constituted by the old stock, to which is added the inputs, minus the outputs. There are then different methods find calculations. The difficulty lies in finding at a given moment the value of the stock, given that the new products are bought at a price a and the old products are bought at a price b, but which are devalued. Several methods exist for calculating the value of the stock, including:

– the weighted average cost per unit (AVCO);

– the cost per batch;

– the last known price;

– the standard cost.

The AVCO is calculated over a fixed period. We distinguish the case of purchased stocks from that of manufactured stocks. In the first case, we have:

$$\text{AVCO} = \frac{\text{Value of the stock at the start of the period + value of the inputs}}{\text{Volume of the stock at the start of the period + volume of the inputs}}$$

and for manufactured stocks:

$$\text{AVCO} = \frac{\text{Value of the stock at the beginning of the period + production cost}}{\text{Quantity in stock at the beginning of the period + quantity produced}}$$

For the batch cost, there are two methods:

– FIFO: (*first in first out*) releases the oldest items.

– LIFO: (*last in first out*) applies the price closest to the current market.

For these two variants, it is necessary to retain the price at which each item was purchased. These methods impose the separate management of different batches input into a stock. Valuation may also be done by the last known price, i.e. the price of the last item purchased. Finally, the manager may decide to set an arbitrary price (the mean for the period, for example), which will be the standard cost for all of the items. Note that all these evaluation methods do not give the same result, and we may have deviations of more than 15%. Table 4.6 shows the different valuation methods on a sample of 12 values. We see that there are deviations at the end of the period. The value of the stock varies from €105 to €122, i.e. nearly 14%. Now that the managers can know the value of the stock, they will wish to optimize their management.

Date	Inputs	Value (€)	Outputs	Vol.	AVCO (€)	FIFO (€)	LIFO (€)	Std 1.60 (€)	Last known price (€)
Jan.				100	150	150	150	160	150
Feb.	300	1.56		400	618	618	618	640	624
Mar.			80	320	494	498	493	512	499
Apr.			140	180	278	280	274	288	280
May	150	1.60		330	518	520	514	528	528
June			130	200	314	318	306	320	320
July			110	90	141	144	135	144	144
Aug.	150	1.70		240	396	399	390	384	408
Sept.			140	100	165	170	152	160	170
Oct.	200	1.75		300	515	520	502	480	525
Nov.			150	150	257	262	239	240	262
Dec.			80	70	120	122	105	112	122

Table 4.6. *A stock valuation*

4.3.5. *ABC classification method*

Before any optimization, it is necessary to effectively target the products upon which the manager wishes to act. It is unrealistic to wish to optimize everything, as the effort required is often too significant compared to the expected gain. It is more profitable to expect to save 10% on items that generate 80% of the costs than on items that generate only 5%. To effectively target the items to be studied, several classification methods exist, among which the best-known method is that of Pareto, or the ABC or 80-20 method. This method classifies items according to their importance for the business. It consists of differentiating the items into three classes (A, B or C) according to a criterion. The choice of the criterion is very important because it determines the optimization strategy to be implemented. As a general rule, the classes correspond to the following percentages:

– Class A: items in small numbers (10–20%) representing a large percentage of the criterion (70–80%).

– Class B: items in medium numbers (10–20%) representing a medium value of the criterion (15–28%).

– Class C: items in large numbers (40–70%) representing a small value of the criterion (2–5%).

We may also choose to use a D class for the items that lie outside the analysis.

In the case of stock management, and according to the aim of the manager, the criteria for this ABC classification may be the following:

– the periodic output values of the stock (year, month);

– the values in stock (average stock);

– the periodic output amount of the stock.

The procedure of ABC classification is the following:

– choose the criterion;

– collect the information;

– classify the items in decreasing order of the criterion;

– calculate the percentage of each item;

– calculate the cumulative percentage of the values of the criterion;

– calculate the cumulative percentage of the number of items;

– draw the curve with the criterion on the y-axis and the items on the x-axis;

– calculate the discrimination ratio;

– obtain the classification according to this ratio.

The discrimination ratio, written DR, lets us determine the item classes. It is determined directly from the ABC curve in the following way:

$$DR = \frac{LM}{KM} \qquad [4.51]$$

An example is given in Figure 4.2. The value of this discrimination coefficient directly gives us the percentages for each class A, B and C. These values are given in Table 4.7. There are two extreme cases for Pareto curves. In the case of the α line, where all of the products have the same percentage, we cannot make a classification. In the case of the β line, where a single item makes up 100% of the criterion, it is useless to consider the other items as they are negligible for the chosen criterion.

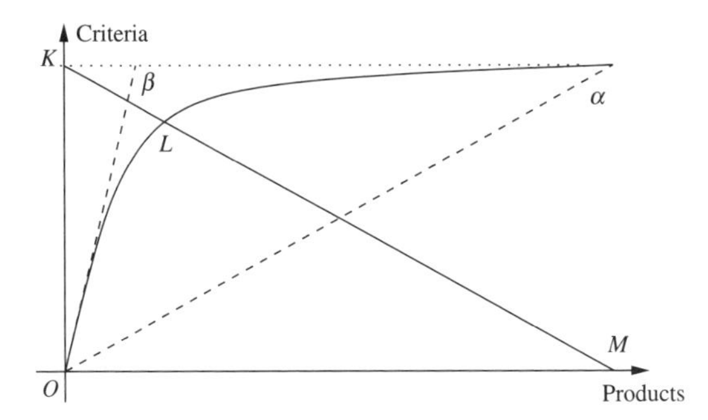

Figure 4.2. *An example of a Pareto curve with the discrimination line*

DR	Percentage of items		
	A	B	C
$1 > DR \geq 0.9$	10	10	80
$0.9 > DR \geq 0.85$	10	20	70
$0.85 > DR \geq 0.75$	20	20	60
$0.75 > DR \geq 0.65$	20	30	50
$0.65 > DR$	Not interpretable		

Table 4.7. *The discrimination ratio*

After classifying the different stocked products according to their significance for a certain criterion, it is necessary to implement a stock management policy, also known as the replenishment method. The first question that the manager then asks is when and how much to order. In a context of simplified demand and delivery, we will first respond to the question of how much to order. The question of when to order may be resolved by setting a periodic order date.

4.3.6. *Economic quantities*

First, we suppose that the quantity of products ordered by the customer is constant and the delivery times are either stable or unstable. Then, the model will be extended to the case where we have a variation of ordering and purchasing costs (reduction), then take into account shortfalls (stockouts), and finally continuous replenishment.

The latter case is found in production facilities in the supply between two machines. In all of these cases, revision of the stock is continual (no inventory period) and the triggering of an order is done as soon as the stock reaches a certain level (zero in the first case described).

4.3.6.1. *Economic quantity: the Wilson formula*

This method is also called the "Economic Order Quantity" (EOQ) method. The approach is based on the Wilson formula. It is characterized by the fact that the size of the ordered batch is not necessarily adjusted to the demand, and this may lead to the risk of residual stock that, of course, has an impact on costs. This method is applied when:

– the demand is relatively constant;

– the purchased items or products are in discrete batches;

– the preparation and possession costs, as well as the delays, are constant and known;

– reordering is done at the same time.

To calculate this economic quantity, it is necessary to calculate the total storage cost. The total storage cost, written as C_T, is the sum of the acquisition costs of the products C_a (the cost of making an order) and the holding costs C_d (possession). This total cost does not take into account the purchasing cost of the products:

$$C_T = C_a + C_d \qquad [4.52]$$

where

$$C_a = a \times n = a \times \frac{D}{Q} \quad \text{and} \quad C_d = T_d \times u \times \frac{Q}{2} \qquad [4.53]$$

where

– a is the acquisition cost (in €) per order;

– $n = \frac{D}{Q}$ is the number of orders;

– D is the predicted annual demand (total quantity of items);

– Q is the quantity ordered per order;

– $\frac{Q}{2}$ represents the average (annual) stock (see Figure 4.1);

– T_d is the holding rate (% of the price of the average stock);

– u is the unit cost of the product.

The economic quantity Q_e is obtained by taking the derivative of the total costs by the quantity $\frac{dC_t}{dQ} = 0$. Thus, we obtain:

$$Q_e = \sqrt{\frac{2 \times a \times D}{T_d \times u}} \qquad [4.54]$$

Figure 4.3 gives a representation of the different costs for obtaining the economic quantity. In the same way, we obtain the optimal number of orders n_e by taking the derivative of the total cost by the number of orders $\frac{dC_t}{dn} = 0$. We obtain:

$$n_e = \sqrt{\frac{T_d \times D \times U}{2 \times a}} \qquad [4.55]$$

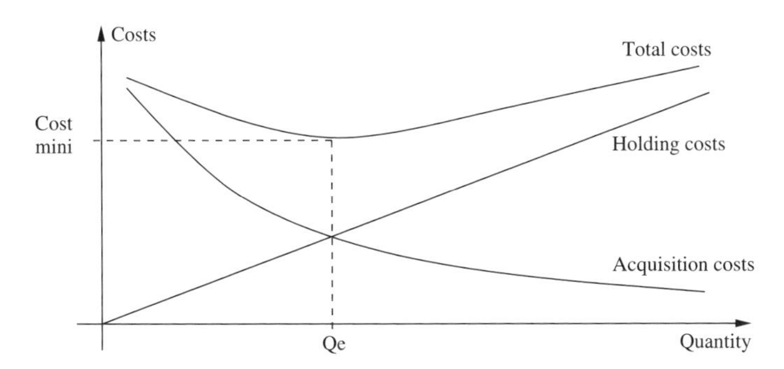

Figure 4.3. *Calculation of the economic quantity*

Hence, we have deduce the optimal order periodicity p_e in months with $p = \frac{12}{n}$:

$$p_e = \sqrt{\frac{288 \times a}{T_d \times D \times u}} \qquad [4.56]$$

EXAMPLE 4.8.– Suppose that the demand D is 2,000 items for the year, with a unit price of $u =$ €2.35. The holding rate T_d is 15% and the cost of placing an order a is €0.5. We wish to determine the economic quantity (number of products per order), the number of orders to be placed per year and the order period. With the previous formulas, we obtain $q_e = 76$ items per order, $n = 27$ orders per year and the period is $p = 13$ days.

4.3.6.2. *Economic quantity with a discount threshold*

At present, we suppose that we have a discount r that is a function of the purchased quantity. To find the optimal quantity to order, we must compare the purchasing budgets with and without a discount. The purchasing budget, written as B, is the sum of the total storage costs (costs of placing an order and of possession) and the value of the products (purchasing cost):

$$B = T_d.u.\frac{Q}{2} + a.\frac{D}{Q} + D.u \qquad [4.57]$$

where $D.u$ is the supplier revenue and $Q = \sqrt{\frac{2 \times a \times D}{T_d \times u}}$.

The budget with the discount, written B_r, is then equal to:

$$B_r = T_d.u'.\frac{Q'}{2} + a.\frac{D}{Q'} + D.u' \qquad [4.58]$$

where $u' = u \times (1 - r)$ (in%) and $Q' = \sqrt{\frac{2 \times a \times D}{T_d \times u'}}$.

It then suffices to compare the budgets B and B_r; if $B_r < B$, then $Q_e = Q'$, and if $B_r > B$, then $Q_e = Q$.

EXAMPLE 4.9.– We take again the previous example with $D = 2{,}000$ items for the year, a unit price of $u =$ €2.35, a holding rate T_d of 15% and an order placement cost a at €0.5. The discount is 5% for every order of more than 100 items. We then seek the quantity to order (with or without the discount). Applying the previous formulas, we obtain $Q = 76$ items and $B = 4{,}726$ without the discount and $q = 100$ and $B_r = 4{,}492$ with discount. Therefore, we choose the solution with the discount.

4.3.6.3. *Economic quantity with a uniform discount*

In this case, we consider the unit price u_i to be dependent on the quantity Q_i ordered. We have a single price for a portion of quantity $Q_i \in [b_{i-1}, b_i[$. For each portion, we calculate the quantity $Q_i = \sqrt{\frac{2 \times a \times D}{T_d \times u_i}}$:

– if $Q_i \in [b_{i-1}, b_i[$, then $\overline{Q_i} = Q_i$;

– if $Q_i < b_{i-1}$, then $\overline{Q_i} = b_{i-1}$;

– if $Q_i > b_i$, then $\overline{Q_i} = b_i - 1$.

Next, we compare the different total budgets B_i:

$$B_i = T_d.u_i.\frac{\overline{Q_i}}{2} + a.\frac{D}{\overline{Q_i}} + S.u_i \qquad [4.59]$$

The choice of $\overline{Q_i}$ corresponds to the minimal B_i.

EXAMPLE 4.10.– Let us consider a demand of $D = 20{,}000$ items/year, a holding rate of $T_d = 15\%$, and an order placement cost of €1 per unit. The unit prices per portion are the following:

– $u_1 = €10$ if $Q < 100$;

– $u_2 = €9.5$ if $100 \leq Q < 200$;

– $u_3 = €9$ if $Q \geq 200$.

The following table gives us the different calculated costs and total budgets.

u_i	Q_i	$\overline{Q_i}$(€)	C_{Ti} (€)	B_i (€)
10	163	99	276	200.276
9.5	167	167	238	190.239
9	172	200	235	180.235

The choice will therefore be made for a quantity of 200 items. Thus, the total budget amounts to €180,235.

4.3.6.4. *Economic quantity with a progressive discount*

In the case of a progressive discount, the first units have a high price, and the following units have a decreasing price. For each portion $[b_{i-1}; b_i[$, it is necessary to calculate $Q_i = \sqrt{\frac{2 \times a \times D}{T_d \times \overline{u_i}}}$, where $\overline{u_i} = u_i + \frac{\sum_{j=1}^{i} b_{i-1}(u_{j-1} - u_j)}{Q_i}$. Next, we choose the quantity $\overline{Q_i}$ that gives the minimum budget:

– if $Q_i \in [b_{i-1}, b_i[$, then $\overline{Q_i} = Q_i$;

– if $Q_i < b_{i-1}$, then $\overline{Q_i} = b_{i-1}$;

– if $Q_i > b_i$, then $\overline{Q_i} = b_i - 1$.

We then calculate each budget: $B_i = T_d.\overline{u_i}.\frac{\overline{Q_i}}{2} + a.\frac{D}{\overline{Q_i}} + D.\overline{u_i}$. We make the choice of $\overline{Q_i}$ corresponding to the minimal B_i.

EXAMPLE 4.11.– Let us consider an annual demand of $D = 20{,}000$ items, with a holding rate of $T_d = 15\%$ and a unit ordering cost of $a = €1$. The unit portion prices are the following:

– $u_1 = €10$ if $Q \in [0; 100[$;

– $u_2 = €9.5$ if $Q \in [100; 200[$;

– $u_3 = €9$ if $Q \geq 200$.

The main difficulty in the resolution of this problem is that the mean unit cost $\overline{u_i}$ is calculated as a function of the economic quantity $\overline{Q_i}$, and this same economic quantity $\overline{Q_i}$ is calculated as a function of $\overline{u_i}$. This difficulty mainly appears in intermediate portions. To address this difficulty, and only for the intermediate portion, we have taken the value of the economic quantity calculated in the first portion, i.e. $Q = 163$. The following table gives the different values of the total budgets B_i as a function of the different portions.

Q	$\overline{u_i}$	Q_i	$\overline{Q_i}$	B_i (€)
$Q < 100$	10	163	99	200.276
$Q = 163$	9.80	165	165	196.316
$Q = 200$	9.75	165	200	195.246

The choice will then be made of an economic quantity of 200 items for a minimum budget of €195.246.

4.3.6.5. *Economic quantity with a variable ordering cost*

In this case, the cost of the order is dependent on the quantity. Indeed, it may integrate administrative and transport costs. It is then often preferable to increase the amount ordered to decrease the order cost. The calculation of the economic quantity follows the same procedure as before. First, we must calculate for each portion $[b_{i-1}; b_i[$ the value of Q_i with $Q_i = \sqrt{\frac{2.a_i.D}{T_d.u}}$:

– if $Q_i \in [b_{i-1}, b_i[$, then $\overline{Q_i} = Q_i$;

– if $Q_i < b_{i-1}$, then $\overline{Q_i} = b_{i-1}$;

– if $Q_i > b_i$, then $\overline{Q_i} = b_i - 1$.

We must then evaluate each budget $B_i = T_d.u.\frac{\overline{Q_i}}{2} + a_i.\frac{D}{\overline{Q_i}} + D.u$ and the choice will be made of the $\overline{Q_i}$ that gives a minimal budget B_i.

EXAMPLE 4.12.– Let us consider the case where the demand D is for 2,000 items/year, the holding rate T_d = 15% and the unit cost u = €1. The unit order placement costs are the following:

– $a_1 = €10$ if $Q \in [0; 400[$;

– $a_2 = €15$ if $Q \in [400; +\infty[$.

We obtain the following results:

a_i	Q_i	$\overline{Q_i}$	B_i (€)
$a_1 = 10$	516	399	80
$a_2 = 15$	632	632	95

The minimal budget is therefore obtained with a unit price of €10. This gives an order quantity of $\overline{Q_1} = 399$ items.

4.3.6.6. *Economic quantity with order consolidation*

Here, the principle is to group together the orders of different items from the same provider. This consolidation may, in certain cases, allow us to obtain reductions in various costs (ordering, transport, etc.). The calculation of the number of orders n to make for k items is as follows:

$$n = \sqrt{\frac{T_d.\sum_{i=1}^{k} D_i.u_i}{2.a.k}} \quad \text{and} \quad Q_i = \frac{D_i}{n} \qquad [4.60]$$

where

– k: the number of types of item per order;

– D_i: the annual demand for product i;

– u_i: the unit cost per product i;

– a: the order placement cost ($\forall i \quad a_i = a$).

To justify the consolidation, we should compare the total storage costs with and without consolidation and take the least expensive solution. Without consolidation, the total cost is:

$$C_t = \sum_{i=1}^{k} C_{ti} = \sum_{i=1}^{k} (T_d.u_i.\frac{Q_i}{2} + a.\frac{D_i}{Q_i}) \qquad [4.61]$$

With consolidation, we save on order placement costs, and we obtain:

$$C_t = a.n + \sum_{i=1}^{k} T_d.u_i.\frac{Q_i}{2} \qquad [4.62]$$

EXAMPLE 4.13.– We will consider three items. Item 1 has an annual demand D_1 of 500 items, with a unit price of $u_1 = 100$. Item 2 has an annual demand of 1,000, with a unit cost of $u_2 = 30$. Item 3 has a demand of $D_3 = 1{,}500$ and a unit cost of $u_3 = 150$. All three items have the same holding rate $T_d = 25\%$ and the same unit order placement cost $a = 90$.

Without consolidation, we obtain:

– $Q_1 = 60$, $n_1 = 8$;

– $Q_2 = 155$, $n_2 = 6$;

– $Q_3 = 85$, $n_3 = 18$.

Therefore, the total storage cost is $C_t = \sum_{i=1}^{3} C_{ti} = 1.500 + 1.162 + 3.182 =$ €5.844.

In the case with consolidation, we obtain:

– $Q_1 = 42, n = 12;$

– $Q_2 = 83, n = 12;$

– $Q_3 = 125, n = 12;$

– total storage cost C_t = €5.844.

Therefore, we have only one order each month ($n = 12$ orders per year) with altered economic quantities. The total cost with consolidation is less than that without consolidation. The choice of consolidating orders is therefore economically justified.

4.3.6.7. *Economic quantity with a non-zero delivery time*

In the case where the delivery time is non-zero, it is necessary to cover the demand during the replenishment period. This period corresponds to the time between the date of order placement and the delivery date. We may then decide to either order a greater quantity, or to anticipate the order so that it is delivered before the stockout. If we do not wish to change the EOQ, we must then calculate the moment to order, and not simply wait for the stock to be zero. Since the stock is not constantly monitored, the order will be made as soon as the stock reaches a level r, which is calculated with the following formula:

$$r = D \times L \qquad [4.63]$$

where

– D is the daily demand.

– L is the delivery time (in days).

Of course, we must verify that the delivery time is less than the time between deliveries.

EXAMPLE 4.14.– Let us consider the case where the annual demand D is for 2,000 items and the delivery time L is 2 days. We then have $D = 5.48$ items/day and $r = 11$ items. Thus, an order must be placed as soon as the level of stock reaches 11 items.

4.3.6.8. *Economic quantity with progressive input*

The notion of progressive input corresponds to the fact that the replenishment of stock is done in a continuous manner, and not in batches, as in the cases considered previously. This configuration is found in production units where an upstream machine supplies the buffer of a downstream machine piece by piece. Figure 4.4 shows the evolution of the stock of the downstream machine. In the first phase, the stock is supplied by the upstream machine with the production of F pieces per hour, while the downstream machine consumes D pieces per hour. Of course, to avoid the emptying of the downstream machine, $F > D$. Therefore, the slope is $F - D$. In the second phase, the upstream machine does not replenish the downstream machine, and there is therefore a consumption of D pieces per hour (a slope of $-D$). The economic quantity is then calculated in the following way:

$$Q_e = \sqrt{\frac{2.a.D.F}{T_d.u.(F-D)}} = \sqrt{\frac{2.a.D}{T_d.u}} \times \sqrt{\frac{F}{F-D}} \qquad [4.64]$$

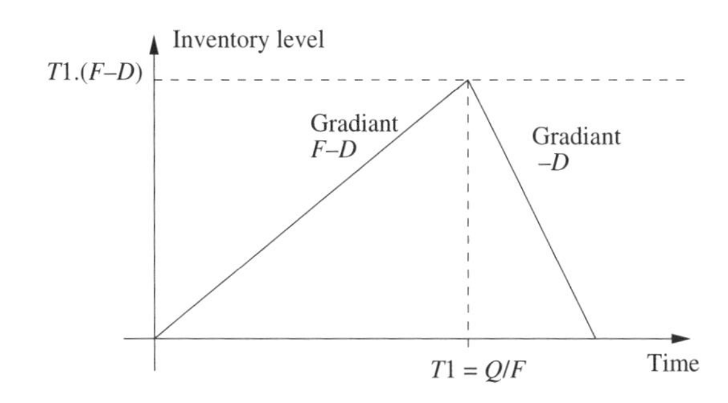

Figure 4.4. *The evolution of the stock with continuous replenishment*

EXAMPLE 4.15.– We consider the annual demand D of the upstream machine to be 8,000 pieces, with a production launch cost of $a = 200$ (the cost of placing an order), an annual production by the upstream machine of F = 12,000 items, a holding rate $T_d = 0.25$, a unit price u = €2,000. We then obtain the economic quantity $Q_e = 139$.

4.3.6.9. *Economic quantity with tolerated shortage*

In this case, stockouts are allowed. We consider the context where the customer demand is deferred but the sale is not lost. We then introduce the stockout cost to take the shortfall into account. Figure 4.5 shows the evolution of a stock where stockouts are tolerated. This shortfall is indicated in gray. The economic quantity is given by the following formula:

$$Q_e = \sqrt{\frac{2.a.D}{T_d.u}} \times \sqrt{\frac{T_d.u + C_r}{C_r}} \qquad [4.65]$$

and the resupply level:

$$N_r = \sqrt{\frac{2.a.D}{T_d.u}} \times \sqrt{\frac{C_r}{T_d.u + C_r}} \qquad [4.66]$$

where:

– C_r: is the stockout cost.

– $T_d.u$: is the holding cost of an item.

– N_r: is the replenishment level.

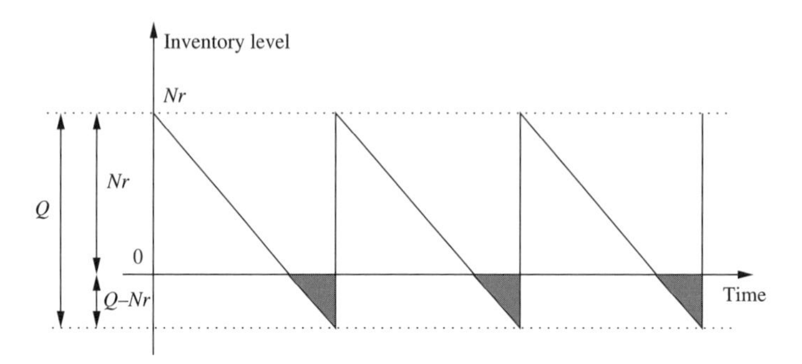

Figure 4.5. *The evolution of the stock with tolerated shortage*

EXAMPLE 4.16.– We consider a demand D = 8,000, an order cost a = €20, a stockout cost C_r = €300, a holding rate T_d = 25% and a unit cost u = €200. We then obtain an economic quantity $Q_e = 86$ and a replenishment level $N_r = 74$.

We have defined some simple rules for calculating the EOQ, we will now describe the different replenishment methods.

4.3.7. *Replenishment methods*

When we analyze the evolution of a stock, we see that the external parameters that determine the evolution are the customer demand and replenishment times. Hence, the replenishment method must answer the question of when and how much to order. To answer this question, replenishment or resupply methods will be characterized by:

– the quantity to be ordered: fixed or variable;

– the review period of the stock: continual or periodic;

– the threshold, or level, below which the manager will make an order.

Stock-taking may be done visually in the physical stock, which is most often a periodic review, but may be continual with stock management software that will automatically signal as soon as the stock is below a certain level. Note that the order is not automatically made. In the first case, the review is of the real stock, whereas this is done on the computerized stock in the second case. The following figures show the difference between the stock position and the physical stock.

4.3.7.1. *The (r, Q) replenishment method*

This method is characterized by the continuous monitoring of the stock (continual review) and the ordered quantity is fixed. The operation is as follows: as soon as the position of the stock falls below the level r, known as the ordering point, a fixed quantity Q is ordered. The order is received after a delivery time L. Figure 4.6 gives an example of replenishment using the (r, Q) method. It is difficult to consolidate orders as they are not triggered at the same time. Stock monitoring is computerized. The quantity Q corresponds to the economic quantity (Wilson formula) and the level r is calculated using the formula given in section 4.3.6.7.

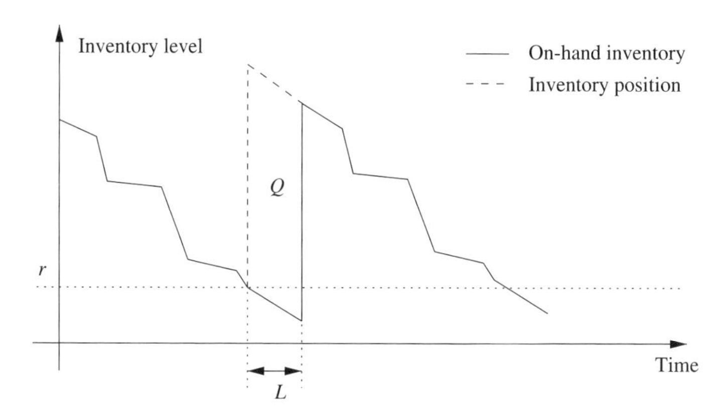

Figure 4.6. *The (r, Q) replenishment method*

4.3.7.2. *The (T, S) replenishment method*

In this method, stock monitoring is periodic, with period T, and the ordered quantity is variable. The procedure is as follows: at each period T, we make an order to reach the replenishment level S. The order is then received after a delivery time L. This method allows the consolidation of the orders of several items at the review times T, $2T$. However, this method is "blind" during the review periods. If demand is too high between these periods, then there is a shortage of stock. Replenishment is done in small quantities. Figure 4.7 gives an example evolution. We have variable order quantities ($Q1$, $Q2$).

4.3.7.3. *The (s, S) replenishment method*

In this method, stock monitoring is continuous, review is continual and the ordered quantity is variable. As soon as the position of the stock falls below an order threshold s, it is replenished up to the replenishment threshold S. The ordered quantity is fixed if demands are unitary, $Q = S - s$. On the other hand, it is variable if demands are made in batches. The ordered quantity is then $Q = S-$the amount of stock when it falls below s. Figure 4.8 gives an example evolution.

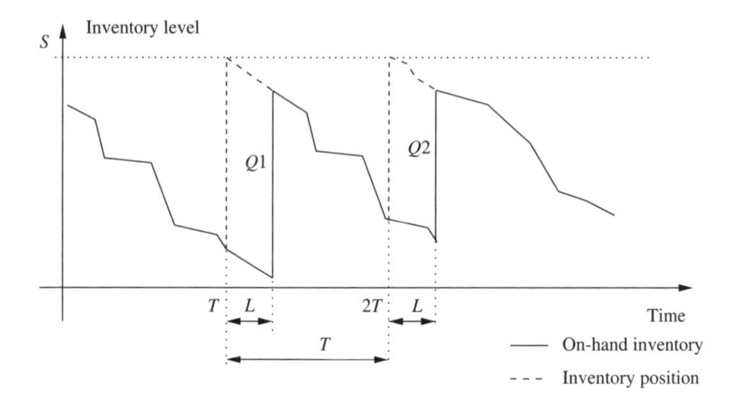

Figure 4.7. *The (T, S) replenishment method*

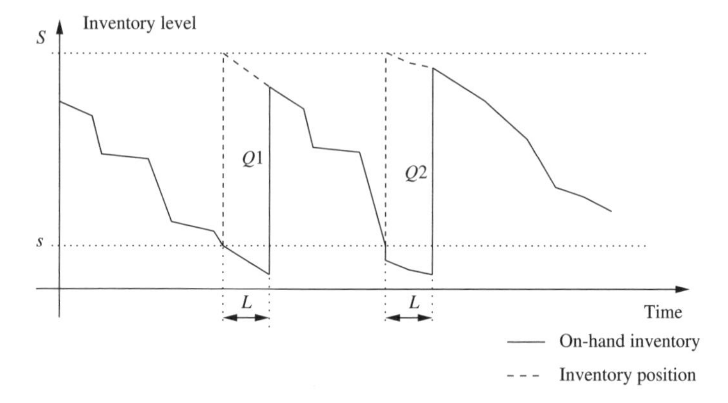

Figure 4.8. *The (s, S) replenishment method*

4.3.7.4. *The (T, r, S) replenishment method*

In this method, stock-taking is periodic, with period T, and the ordered quantity is variable. At the end of each monitoring period T, we examine the level of the stock and only place an order if it is below r, thus returning the level of stock to the replenishment threshold S. Figure 4.9 gives an example evolution. We have variable order quantities $Q1$, Q_2 at times T, $2T$.

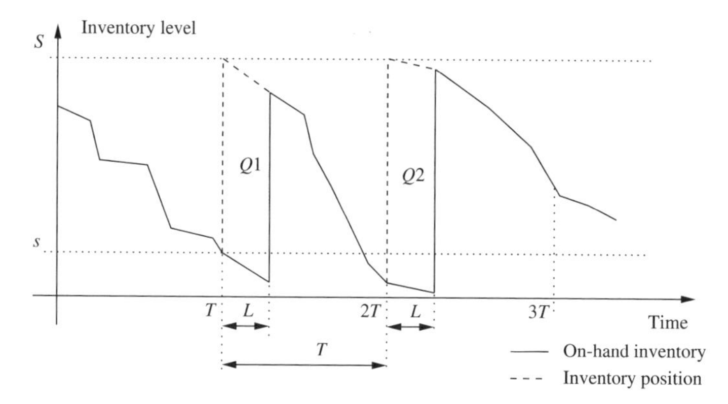

Figure 4.9. *The (T, r, S) replenishment method*

4.3.7.5. *The (T, r, Q) replenishment method*

In this method, stock monitoring is periodic, with period T. This method is a combination of the (r, Q) and (T, S) methods. Thus, at the end of each period T, if the position of the stock is below the order point r, then a fixed quantity Q is ordered. This method avoids making orders of too small a size. Figure 4.10 shows an example of the evolution of a stock with this method.

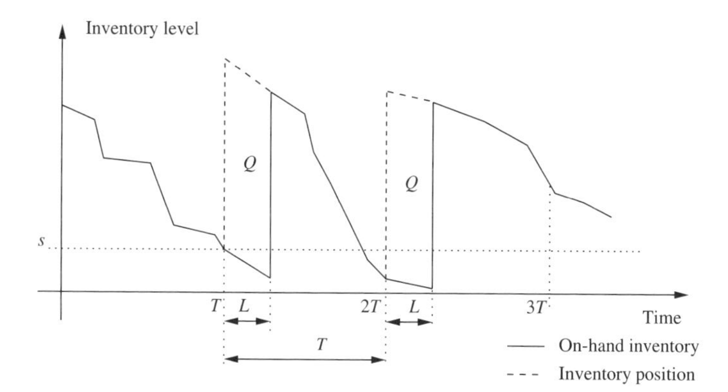

Figure 4.10. *The (T, r, Q) replenishment method*

4.3.7.6. *Security stock*

All of the replenishment methods presented earlier have the main goal of having the minimum storage cost that maintains a service rate to the customer. However, it may occur that the demand D is greater than that anticipated, and the stock is then short of items. This risk is more significant during the delivery period, as the stock level is below the order point r and s. To overcome this problem, the manager will often make use of a safety stock that can absorb this exceptional demand. The security stock is therefore a dormant stock, which is normally very rarely used. In the case of perishable products, it is necessary to perform a rotation of the items to avoid having perished, and therefore useless, products in the stock. The determination of the security stock level is carried out using a calculation based on consumption statistics, i.e. from means and standard deviations. This involves an admissible stockout rate, which corresponds to the risk that the manager agrees to take. The level of the security stock SS is given by the following formula:

$$SS = k.\sigma.\sqrt{L} \qquad [4.67]$$

where:

– k is the stockout coefficient. This is the risk accepted by the manager.

– σ is the standard deviation (calculated over the consumption history).

– L is the lead time.

We note that the larger k is, the smaller the risk and the more the security stock is important. The following table gives the k coefficient as a function of the accepted risk. For any risk other than those given in the table, the k coefficient may be read from a table of the normal distribution.

Risk	16%	5%	2.5%	1%	0.5%
k	1	1.65	1.95	2.35	2.58

Depending on the replenishment method used, the lead time may change:

– For order point methods: lead time = L.

– For calendar methods: lead time = $L + T$.

EXAMPLE 4.17.– We consider the case where σ = 50 items (the standard deviation of monthly consumption over a year) and L = 2 weeks. The enterprise takes a risk of 1% (from the normal distribution). We then obtain a stockout coefficient of k = 2.35 and a security stock $SS = 2.35 \times 50 \times \sqrt{0}.5 = 83$ items.

Figure 4.11 gives an example of stock evolution with a security stock.

4.4. Cutting and packing problems

Once the most appropriate stock management policy for the problem is determined, we should consider the optimization of the warehousing of the stocked objects. This problem is described in detail in the work of Amodeo and Yalaoui

[AMO 05]. When dealing with warehousing, we come across *cutting and packing* problems. In fact, we encounter them in numerous industrial contexts when it comes to the minimization of wasted raw materials from cutting operations (cutting plastic, paper, metal, etc.) or the maximization of the number of objects can be stored in a defined area.

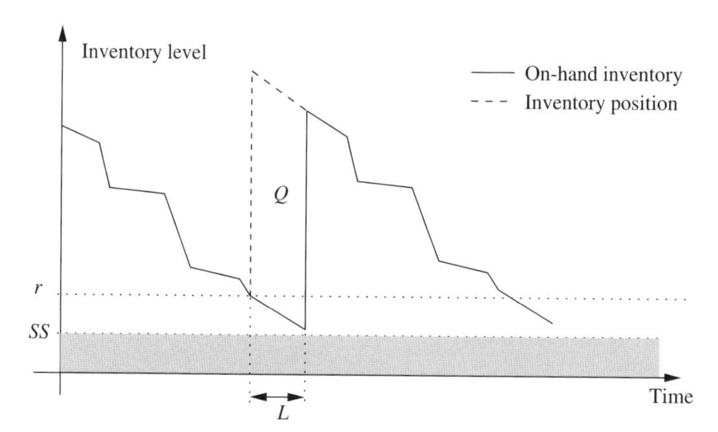

Figure 4.11. *An example of a security stock*

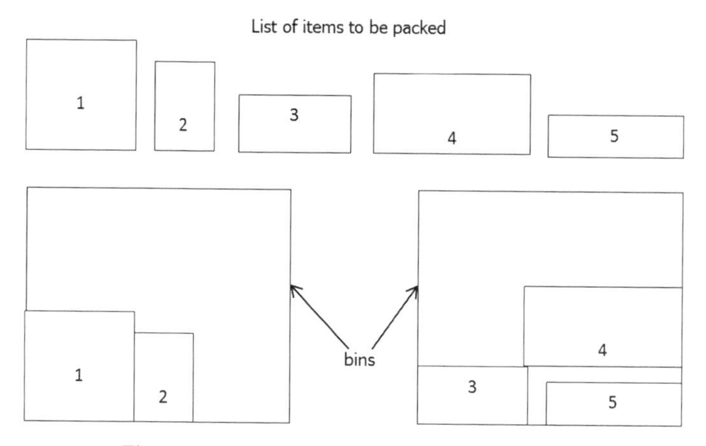

Figure 4.12. *Packing rectangular-shaped pieces*

4.4.1. *Classifying cutting and packing problems*

There are many variants of these problems, which leads to the creation of different classifications depending on the constraints, the function that we seek to maximize and the number of considered dimensions. The problem may be one-dimensional (we must cut a bar of metal into different pieces and we seek, for example, to minimize waste), two-dimensional (we wish to arrange some rectangular boxes in a storage area which is itself rectangular, without stacking them, so that we maximize the number of boxes that we can store there) or three-dimensional (we must fill a truck with rectangular objects that we may stack).

The first classification was proposed by Dyckhoff in the 1990s [DYC 90]. It is based on four characteristics that provide information about the dimensionality of the problem, the kind of assignment, the assortment of containers (called bins) and the assortment of items to be packed:

– Dimensionality: this characteristic determines the dimensions of the considered items. These objects may be unidimensional or multidimensional. We generally find a dimensionality of 1, 2 or 3.

– Kind of assignment: this second characteristic indicates the type of assignment of the small items to the large objects (containers). We distinguish between two types of allocation, written as *(B)* and *(V)*:

- (B) where all of the bins are used. A subset of items is assigned to these bins. In this case, we seek to maximize the number of items to assign to the available objects;

- (V) where all of the items are assigned to a subset of bins. In this case, we seek to minimize the number of bins.

– Assortment of bins: we distinguish between three cases: a single container, written as O; several identical containers, written as I, and finally several different containers, which we represent by D.

– Assortment of items: Dyckhoff distinguishes between few items of different figures (F), many items of different figures (M), many items of relatively few figures (R) and, finally, congruent figures (C).

Table 4.8 summarizes the different characteristics.

Other classifications have been introduced, such as that of Lodi *et al.* [LOD 04] that, like its predecessors, does not allow the differentiation of every problem encountered, but remains widely used as it identifies most typically encountered two-dimensional problems. It consists of three fields:

– The first field indicates whether the problem is one of *bin packing* (B) or *strip packing* (S). The first case concerns the cutting (or packing) of small objects from (or in) several large objects. In the second case, given a strip of infinite length, we seek to minimize the length used.

– The second field gives information about the assumptions of the problem: the letter "O" indicates that rotation is not permitted (in the case of cutting fabric with patterns, for example) and the letter "R" indicates that it is permitted.

– The last field provides information about the guillotine constraint. This constraint, also known as the end-to-end cutting constraint, comes from the use of cutting tools that require this restraint to be respected. The letter "G" then indicates that the problem has guillotine constraints, and the letter "F" indicates that it does not have (Figure 4.13).

Characteristic	Symbol	Description
Dimension	1	One dimension
	2	Two dimensions
	3	Three dimensions
Type of task	B	A subset of items and a set of bins
	V	A set of items and a set of bins
Assortment of objects of large sizes	O	A single bin
	I	Identical bins
	D	Different bins
Assortment of objects of small sizes	F	Few items of different shapes
	M	Several items of different shapes
	R	Items of relatively identical shapes
	C	Identical items

Table 4.8. *The classification of cutting and packing problems [DYC 90]*

For example, we may encounter the *2BP/O/G* problem in which we must cut small objects out of (two-dimensional) larger ones, without changing their orientation and following the guillotine constraint. Another example is *2SP/R/G*. This time, it is from a strip of raw material of infinite length that we must cut the small objects, the rotation of the small objects being permitted and the guillotine constraint being respected.

Finally, the most recent classification is that of Wäscher *et al.* [WAS 07]. This is the most complete classification; it is based on the division of cutting and packing problems into two large families: problems where we seek to produce (or pack) a maximum of items, given that we may not necessarily be able to satisfy all of the demand (output maximization problems (OMP)) (Figure 4.14), and problems where we seek to satisfy all of the demand while minimizing the resources consumed (input minimization problems (IMP)) (Figure 4.15).

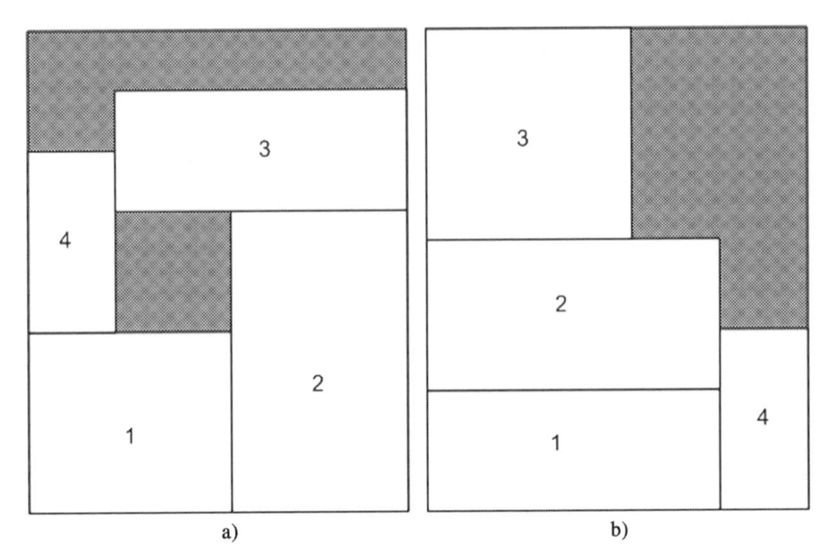

Figure 4.13. *Configurations (a) without and (b) with guillotine restraints*

Characteristics of the large objects \ Assortment of small items		Identical	Weakly heterogeneous	Strongly heterogeneous
Fixed dimensions	A large object	Identical item packing problem IIPP	Single large object packing problem SLOPP	Single knapsack problem SKP
	Identical		Multiple large object packing problem MLOPP	Multiple knapsack problem MKP
	Heterogeneous		Multiple heterogeneous large object packing problem MHLOPP	Multiple heterogeneous knapsack problem MHKP

Figure 4.14. *OMP problems according to the Wäscher et al. classification [WAS 07]*

Assortment of small items / Characteristics of large objects		Weakly heterogeneous	Strongly heterogeneous
Fixed dimensions	Identical	Single stock size cutting stock problem SSSCSP	Single bin size bin packing problem SBSBPP
	Weakly heterogeneous	Multiple stock size cutting stock problem MSSCSP	Multiple bin size bin packing problem MBSBPP
	Strongly heterogeneous	Residual cutting stock problem MSSCSP	Residual bin packing problem RBPP
One large object variable dimension(s)		Open-dimensional problem ODP	

Figure 4.15. *IMP problems according to the Wäscher et al. classification*

4.4.2. *Packing problems in industrial systems*

Cutting and packing problems are often NP-hard (Garey and Johnson [GAR 79]). They are very often encountered in industries, whether for the optimization of machining operations and the cutting of resources, or for the management of the storage of raw materials, works in progress or finished products. These problems are also involved when it comes to loading a vehicle for delivery operations, for example. We will consider here a problem that is encountered most often in stock management: the problem of *bin packing* in two dimensions. We propose a mathematical model and a simple method for its solution.

4.4.2.1. *Model*

We consider a set of n objects to be packed in K storage zones on the ground, with width W and depth H. We assume that we cannot, for reasons of fragility and handling, put the objects on top of each other, which reduces the study to two dimensions. We define a Cartesian coordinate system to describe a possible packing. The origin $(0, 0)$ corresponds to the bottom left corner of the zone. The u-axis of abscissas and the v-axis of ordinates correspond to the sides W and H, respectively. The coordinates of the lower left corner of an object $t = 1, \ldots, n$, packed in a zone $k = 1, \ldots, K$, are written as $u_{t,k}$ and $v_{t,k}$. We introduce the variable M, a very large, finite number, and the binary variable y_{tk} that equals 1 if object t is allocated to zone k and 0 otherwise. The variable Y_k equals 1 if zone k is used, and 0 otherwise.

The two-dimensional bin packing model may then be written in the following way:

$$\min \sum_{k=1}^{K} Y_k \qquad [4.68]$$

Under the constraints:

$$\sum_{k=1}^{K} \sum_{t=1}^{n} y_{tk} \leq MY_k \qquad [4.69]$$

$$\sum_{k=1}^{K} y_{tk} = 1 \ \forall t = 1, \ldots, n \qquad [4.70]$$

$$\sum_{t=1}^{n} y_{tk}(w_t \times h_t) \leqslant W \times H \qquad [4.71]$$

$$u_{tk} + w_t \leqslant W + M(1 - y_{tk}) \ \forall t = 1, \ldots, n, \forall k = 1, \ldots, K \qquad [4.72]$$

$$v_{tk} + h_t \leqslant H + M(1 - y_{tk}) \ \forall t = 1, \ldots, n, \forall k = 1, \ldots, K \qquad [4.73]$$

$$\begin{array}{l} u_{tk} + w_t \leqslant u_{t'k} + M(2 - y_{tk} - y_{t'k}) \\ \text{or} \\ u_{t'k} + w_t' \leqslant u_{tk} + M(2 - y_{tk} - y_{t'k}) \\ \text{or} \\ v_{tk} + h_t \leqslant v_{t'k} + M(2 - y_{tk} - y_{t'k}) \\ \text{or} \\ v_{t'k} + h_t' \leqslant v_{tk} + M(2 - y_{tk} - y_{t'k}) \\ \forall t = 1, \ldots, n, \forall t' = 1, \ldots, m, t \neq t', \forall k = 1, \ldots, K \end{array} \qquad [4.74]$$

$$y_{tk} \in \{0, 1\} \ \forall t = 1, \ldots, n, \forall k = 1, \ldots, K \qquad [4.75]$$

$$y_k \in \{0, 1\} \ \forall k = 1, \ldots, K \qquad [4.76]$$

The objective function [4.68] minimizes the number of zones used. Constraint [4.69] lets us know that a zone is used from the moment it has at least one packed object. Constraint [4.70] ensures that all of the packages are packed only once in a zone. Constraint [4.71] ensures that the area of the set of objects allocated to a zone does not exceed the area of the zone. Constraints [4.72] and [4.73] ensure that the objects do not exceed the limits of the zone. Constraint [4.74] ensures that the objects do not overlap.

4.4.2.2. *Solution*

Numerous methods have been developed to solve bin packing problems, including simple heuristics and metaheuristics, as well as exact branch-and-bound-type procedures [WAS 07]. Here, we present the basic heuristic, i.e. the *bottom left* heuristic.

4.4.2.2.1. The bottom left heuristc

Let us consider the case where the rotation of items is not allowed. In this heuristic, the items are first sorted by the descreasing order of their height. Considering them in this order, we seek to pack them furthest to the bottom left of the zone (bin) during filling, without overlapping them with other items and such that they do not exceed the limit of the zone. If such a packing does not exist, we pack the item in a new zone.

4.4.2.2.2. Example

Consider the following problem with 10 items to be packed in bins with dimensions 15×25.

i	1	2	3	4	5	6	7	8	9	10
h_i	8	11	10	4	12	11	12	3	4	3
w_i	6	4	3	7	6	6	8	6	3	8

Table 4.9. *An example of the bottom left heuristic*

We rank the items in descending order of their height, i.e. in the order 7-5-6-2-3-1-4-9-10-8. We then pack them and obtain the packing given in Figure 4.16.

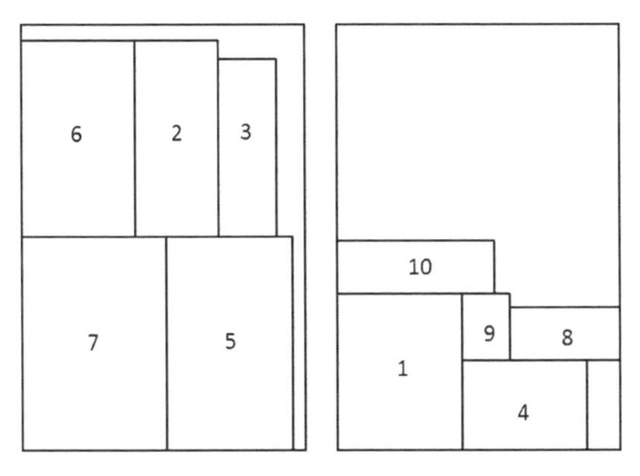

Figure 4.16. *An example of the bottom left heuristic*

4.5. Production and replenishment planning, lot-sizing methods

4.5.1. *Introduction*

From demand forecasts, the workload over the considered time frame is evaluated. This consists of determining the type(s) of product, which will be manufactured for each resource in each period, as a function of the characteristics of the production system. These forecasts allow the planing, for example, of the human resources needed, as well as the raw materials or components, and therefore the prediction of the necessary replenishments sufficiently in advance, so that stockouts are avoided. We therefore end up with the consolidation of requirements by periods – problems known as lot-sizing. These problems are part of the material requirement planning (MRP).

4.5.2. *MRP and lot-sizing*

MRP allows the management of production resources, taking into account their limited capacity. This approach constructs an acceptable solution in three steps. In the first step, from the desired quantity of products, we determine the necessary quantities for the different levels of manufacturing. This is called "explosion" of the bill of materials; it is the calculation of net requirement. This calculation is done ignoring capacity constraints. Then, the second step consists of smoothing out the loads so as to respect capacities while ignoring precedence constraints. Finally, in the third step, we determine the schedule by introducing these constraints.

MRP is used for medium-term planning in push flow management or in repetitive (mass) production. Its goal is to anticipate demand in order to adapt to variations. MRP therefore lets us know when and how much to produce, taking into account stock constraints and problems related to manufacture, with the objective of minimizing costs. To apply MRP into a business, we must make use of an information system, integrating physical, technical and accounting data, and above all the plan of the production manager.

The physical data necessary are, for example, available stocks, expected deliveries, deferred demands, provisional demands, firm orders for raw materials, semi-finished products and finished products. The technical data necessary are, for example, operational, transit and waiting times, bills of materials and production capacities. A macrogamme indicates operational, transit and waiting times. A bill of materials corresponds to the breakdown of the products into components with a repetition factor. A production capacity represents the number of resources available in standard and overtime hours per production center. The accounting data necessary are, for example, manufacturing costs (in normal or degraded operation) and the outstanding values.

The production management plan lists the quantities to be produced per product family, per production center and per period, as well as the stock levels for the raw materials, semi-finished products and finished products per product family. These data are obtained by working groups consisting of the senior management, commercial management and production management teams as well as the finance and personnel departments.

The question of the consolidation of net requirements thus arises. In effect, we must define the size of production or delivery batches, and know how these groupings are introduced. Four ordering techniques (with or without consolidation) are possible:

– The absence of consolidation, meaning that any quantity may be manufactured. The consolidated net requirements are therefore equal to the net requirements.

– The prohibition of quantities that are too small. This technique aims to put forward low demand over one or more periods until a grouping greater than a minimum quantity is obtained. The consolidated net requirements are then equal to the net requirements with the anticipation of low demand.

– *Conditioning* consists of ordering in multiples of economic production or delivery quantities. The consolidated net requirements are equal to the anticipated net requirements consolidated in multiple of these quantities. This technique allows anticipation and consolidation in multiples of economic quantities.

– The variable size of lots adjusted to minimize the related costs (fixed cost of placing an order, cost significant to each order and storage costs). Therefore, it is necessary to dynamically determine the size of lots to minimize the costs on the time frame of a work.

4.5.3. *Lot-sizing methods*

Many lot-sizing models have been developed. In some of them, constraints of resource capacity are not considered, and they lie within the first phase of MRP. Sometimes the assumptions are such that only a single production level and a single type of product are considered (with capacity constraints). The methods proposed for these models then group together the three phases.

To better model the behavior of the system under consideration, it is first necessary to study the elements that will influence decisions. We distinguish, for example, between the transformations to which items are subjected (which result in precedence constraints), the availability of materials and products (the possibility of stockout or the lacking thereof) and resource capacities (capacity constraints). The considered system and decision level determine the choice of parameters, the level of structural detail and the type of constraints, as well as the evaluation criterion.

4.5.3.1. *The characteristic elements of the models*

4.5.3.1.1. The planning time frame and time scale

The planning time frame is the time interval for which hypotheses are made. The demand must be respected therein, and performance is measured using an evaluation criterion. The time frame may be either finite or infinite. The time scale may be continuous (the considered interval is made-up of real numbers) or discrete (the interval is made-up of N integers representing N periods). In the discrete case, the events and decisions that take place in continuous time must be discretized according to the size of the periods constituting the time frame. For example, if small time periods are chosen, it seems reasonable to only allow the manufacture of at most a single item per period.

4.5.3.1.2. The demand

The demand is the input data for lot-sizing models. Two main approaches are studied: one in which the demand is considered to be fixed, and another in which the demand is considered to be dynamic. Random events are not taken into account in the case where it is deterministic, and are modeled by a probabilistic law in the stochastic case.

4.5.3.1.3. The constraints

A constraint is a condition that every viable solution to any problem must satisfy. Among those encountered in lot-sizing problems, the service level is an unavoidable constraint. It is defined as the percentage of orders to be met in a time fixed in advance, as the maximum accepted shortfall. Stock levels and their size require a minimum availability guarantee, known as the security stock. For stochastic demand, a service level of 100% would be costly. Generally, it is expressed as the maximum number of stockouts per year. Therefore, we must specify the conditions under which an order may be backlogged if it is not honored on time.

It is also necessary to take into account limits in resource capacity. If we may integrate the costs of exceeding capacities, or if there are no constraints of this type, the model is described as *uncapacitated*. When these constraints are clearly specified, the model is said to be *capacitated*. Numerous other constraints arising from the diversity of industrial problems are encountered [BRA 06].

4.5.3.1.4. The structure of the production process

The structure of the production process specifies the number of levels through which the items pass. The structure is (*single level*) if all the input materials come from an entity outside the considered system. Other models are (*multi-level*). Some models only consider a single product/item, while others consider multi-products/items.

The structure of the production process also influences the definition of the constraints. Indeed, it may be that limited capacity resources are used by different items. It is then necessary to define rules that set the maximum number of changes of item type (setup) per period.

4.5.3.1.5. The objective function

Generally, the objective is the minimization of a cost per unit of time. The objective function accounts for a cost proportional to the use of capacities, or the level of activity for a given policy. We very often encounter such components as the total cost, holding or storage costs and the setup cost. More rarely, the backlogging cost is considered when this is allowed. We also sometimes encounter the unit purchasing cost while determining a replenishment policy, and this is not fixed. In most models, all costs are considered as deterministic data.

It is important to note that the majority of models developed in the literature are equally applicable for the determination of a production plan as for the determination of replenishment plans.

In effect, it is possible to establish a parallelism between the unit production cost and the unit purchasing cost as well as between the production setup cost and that of order placement. Storage costs are present in both cases. When the discussed problem is one of replenishment or of production planning, it is called a lot-sizing problem.

4.5.3.2. *Lot-sizing in the scientific literature*

Here, the main basic models of lot-sizing are discussed. They are grouped according to the nature of the demand and the presence or absence of capacity constraints.

4.5.3.2.1. Models with stationary demand and without capacity constraints

The EOQ is one of the first models defined to work with lot-sizing [BRA 06]. Numerous models have been developed from this.

We cite, for example the EOQ with joint replenishment cost (EOQJR) model, which adds joint replenishment costs, and the economic order quantity in multi-level-product structure (EOQML) that incorporates EOQ, while considering a more complex structure.

4.5.3.2.2. Models with dynamic demand and without capacity constraints

The optimal solution to this problem may be obtained by using the Wagner–Whitin (WW) algorithm based on dynamic programming. This algorithm was introduced by Wagner and Whitin in 1958 [BRA 06]. Its complexity being $O(n^2)$, where n is the size of the time frame, this algorithm is known to have too great a calculation time for real-sized problems.

To overcome this problem, different heuristics have been implemented in order to find near optimal solutions requiring shorter computational time. We cite, for example the *least unit cost* (LUC) method, the *part period algorithm* (PPA) and the *Silver–Meal* (SM) algorithm.

4.5.3.2.3. Models with dynamic demand and capacity constraints

In 1982, Bitran and Yanasse [BIT 82] discussed the *capacitated lot-sizing problem* (CLSP) an NP-hard problem. This model lends itself particularly well to time frames divided into relatively large periods. One of the extensions of the CLSP is the *multi-level capacitated lot-sizing problem* (MLCLSP), which incorporates the hypotheses of the CLSP, generalizing to the case of a multilevel, multiproduct production structure.

Another model characterized by capacity constraints and dynamic demand is the *continuous setup lot-sizing problem* (CSLP). We have seen that there are several decision levels in production planning. The level of the CLSP is greater than that of the CSLP, as it may be applied to a family of items. The CSLP is applied to a precisely defined item. The characteristics and constraints of a model depend on the chosen level of detail. It is for this reason that the periods constituting the planning time frame of the CSLP are assumed to be relatively small. The capacity for a resource may only be used by a single item per period, and, therefore, there may only be a change of setup at the beginning of each period [BRA 06].

Other variants are also found, such as *discrete lot-sizing and scheduling problem* (DLSP), which includes the same hypotheses as the CSLP with an additional constraint: the capacity available during a period must be either used in its entirety or not at all. This is called the “all or nothing” rule.

Many of these problems are NP-hard in the weak sense, as they may be solved in pseudo-polynomial time. We note that only the following three problems are of polynomial time: when the capacity is infinite with concave functions (Wagner–Whitin models), when the capacity is constant and the functions are concave and in the case where there are no setup costs and the functions are concave in nature, regardless of capacity limits.

4.5.4. *Examples*

We now focus, in particular, on methods used for the *single-product single-level capacitated dynamic lot-sizing model*. The lot-sizing problems grouped under this name are those of a single level, where the resource is limited in capacity and for one product. These problems belong to the CLSP category. We assume that the evolution of the demand over T consecutive periods is known.

The demand in a given period may be satisfied by either the production (or delivery) of the current period, or by the stock established in previous periods. Sometimes, when shortfall situations are allowed, the demand may also be satisfied by the production of future periods. The cost associated with each production or replenishment plan then depends on the production and holding levels in each period on the time frame T.

4.5.4.1. *The Wagner–Whitin method*

The Wagner–Whitin method [WAG 58] is used for problems without capacity constraints. It is based on the analogy of the search for the shortest path on a graph.

The hypotheses are simplistic: we assume that there can be neither delays nor lost orders, and that costs are linear.

4.5.4.1.1. Model

Consider a time frame divided into T periods. x_t is the quantity produced (or ordered) in period $t = 1, \ldots, T$. y_t is a binary variable that equals 1 if there is a production (or order) in period t, and 0 otherwise. We write d_t for the demand in period t, and $p_t(x) = p \times x$ is the production (or order) cost function.

The holding (or storage) cost function is written as $a_t(x) = a \times x$ and, finally, K is the setup (or order placement) cost. The model of the problem is then as follows:

$$\text{Min} \sum_{t=1}^{T} [K \times y_t + p \times x_t + a \sum_{\tau=1}^{t} (x_\tau - d_\tau)] \qquad [4.77]$$

$$\sum_{\tau=1}^{t} (x_\tau - d_\tau) \geq 0 \quad t = 1, 2, \ldots, T \qquad [4.78]$$

$$0 \leq x_t \leq M y_t \quad t = 1, 2, \ldots, T \qquad [4.79]$$

$$y_t \in \{0, 1\} \qquad [4.80]$$

where M is an arbitrary large number.

4.5.4.1.2. Method

The Wagner–Whitin method is based on the search for the shortest path from town 0 to T, is the same as the search for the amounts to cover the demands of periods 0 to T at a minimal cost.

This solution is done by using dynamic programming. For this, we must specify the cost of covering the requirements for periods $i + 1$ to j in a single order (production), which corresponds to going from town i to town j in a single step. This cost, $c_{i,j}$, is the following:

$$c_{i,j} = K + a \sum_{h=i+1}^{j} \sum_{t=h+1}^{j} d_t = K + a \sum_{t=i+2}^{j} d_t(t-i-1) \quad [4.81]$$

We then have the following recursive formulation, where C_i is the smallest cost for arriving at period (town) i:

$$C_0 = 0 \quad [4.82]$$

$$C_h = \min_{0 \le k \le h-1} (C_k + c_{k,h})$$

$$= \min_{0 \le k \le h-1} (K + a \sum_{t=k+2}^{h} d_t(t-k-1)), h = 1, 2, \ldots, T \quad [4.83]$$

$$k^*(h) = \arg\min_{0 \le k \le h-1} (C_k + c_{k,h}) \quad [4.84]$$

4.5.4.1.3. Example of an application

Let us consider the example with six periods in Table 4.10. We have $K = €100$ and $a = €1$ per item per period.

Period t	1	2	3	4	5	6
d_t	70	30	35	60	60	25

Table 4.10. *An example of the Wagner–Whitin method*

We then seek the shortest path between each pair of towns. To go from 0 to 1, we have a single possible path, with cost 100. To go from 0 to 2, we have either a cost of 200 (going through town 1, i.e. covering the requirements for periods 1 and 2 in two production launches), or a cost of 130 (going directly from town 0 to town 2), which corresponds to a single production launch, to which the storage cost of 30 units is added during one period: $100 + 30 = 130$.

The paths between all of the towns are shown in Figure 4.17.

In Figure 4.18, only the shortest path between each pair of towns is shown.

Following this, we seek the shortest path going from town 0 to town 6.

The shortest path has a cost of 410, going through town 3. This means that we place a firm order to cover the requirements of periods 1–3 (delivery at the end of period 0) of 135 units, and we make a provisional order of 145 units to cover the requirements of the other periods.

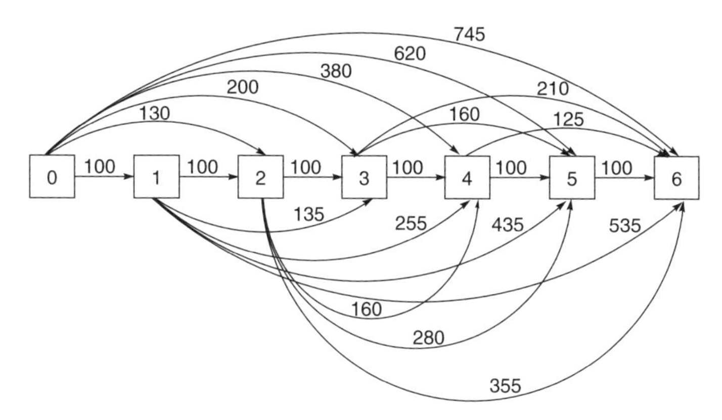

Figure 4.17. *A Wagner–Whitin example*

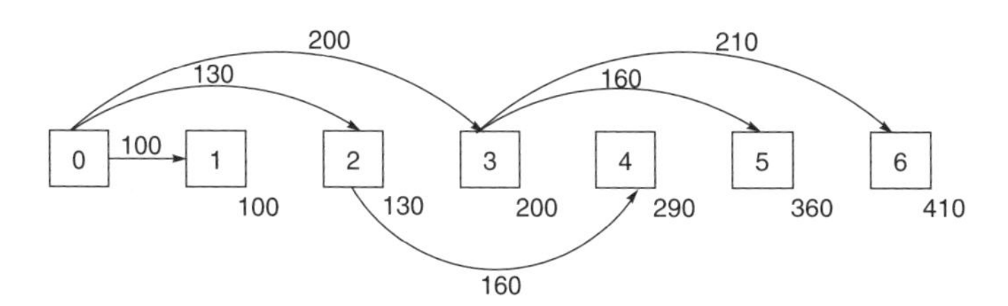

Figure 4.18. *A Wagner–Whitin example*

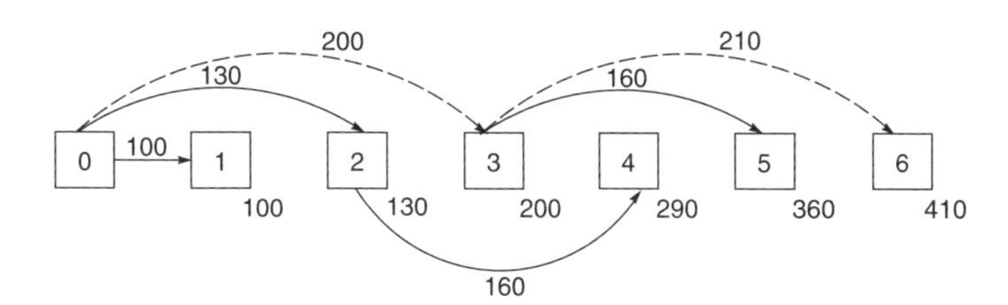

Figure 4.19. *A Wagner–Whitin example*

4.5.4.2. *The Florian and Klein method*

Florian and Klein [FLO 71] studied the case where the maximum capacity available per period is constant across the time frame. They assumed the production and holding cost functions to be concave in nature. They established a decomposition of the problem into independent subproblems, and characterized the extreme points of the closed, bounded and convex set of feasible solutions. From this, they developed a method based on dynamic programming, which they applied to cases with and without shortage.

Their method is divided into two phases: the search for the optimal solution to all of the subproblems and the solution of the overall problem from the results obtained for the subproblems.

4.5.4.2.1. Model

The costs taken into account in this first approach are the holding cost, written as h_i $(i = 1\ldots, T)$, the production cost, written as p_i $(i = 1\ldots, T)$ and the setup cost s_i $(i = 1\ldots, T)$. The quantity produced in period i is written as x_i. For $p'_i(x_i)$, we write the following function:

$$p'_i(x_i) = s_i \times \delta(x_i) + p_i \times x_i \quad [4.85]$$

where

$$\delta(x_i) = 1 \text{ if } x_i > 0, 0 \text{ otherwise} \quad [4.86]$$

The functions $p'_i(x_i)$ and $h_i(I_i)$ are assumed to be concave functions. This means that as the quantities x_i and I_i increase, the production and holding costs per unit decrease. The problem may then be formulated in the following way:

$$\text{Min } \Sigma_{i=1}^{T} p'_i(x_i) + \Sigma_{i=1}^{T} h_i(I_i) \quad [4.87]$$

$$I_i \geq 0 \ \forall \ i = 1, \ldots, T \quad [4.88]$$

$$0 \leq x_i \leq C_i \ \forall \ i = 1, \ldots, T \quad [4.89]$$

$$\sum_{j=1}^{i} C_j \geq \sum_{j=1}^{i} d_j \ \forall \ i = 1, \ldots, T \quad [4.90]$$

$$I_0 = I_T \quad [4.91]$$

where

$$I_i = \sum_{j=1}^{i} x_j - \sum_{j=1}^{i} d_j \ \forall \ i = 1, \ldots, T \quad [4.92]$$

Because the cost functions are concave, so is the objective function. The constraints define a closed, bounded and convex set.

4.5.4.2.2. Method

The Florian and Klein method may be divided into two steps: the solution of the subproblems and that of the general problem.

4.5.4.2.3. Solution of a subproblem

From the assumptions, we have $C_i = C \ \forall\, i = 1, \ldots, T$. The authors showed that the cumulative demand may be decomposed for every subproblem $\overline{S}_{uv}$, such that $I_u = I_v = 0$, in the following way:

$$\sum_{l=u+1}^{v} d_t = kC + \varepsilon, \quad 0 < \varepsilon < C \text{ and } k \text{ is a positive integer} \qquad [4.93]$$

Furthermore, the authors showed that there are at most three possible values for the production level of a subproblem in each period: 0, ε and C. We write as X_j the cumulative production of $u + 1$ to j defined as follows:

$$X_j = \sum_{i=u+1}^{j} x_i \text{ where } j = u+1, \ldots, v \qquad [4.94]$$

This takes its values in the set $\{0, \varepsilon, C, C+\varepsilon, \ldots, kC\varepsilon, kC+\varepsilon\}$. A graph may then be made with the possible values for X_j $(j = u+1, \ldots, v)$ as the vertices and the arcs (X_j, X_{j+1}) defined as follows:

– If $X_j = mC$, with $m = 0,1,\ldots,k$, then the value of the input vertex of the arc is $X_{j+1} = X_j + (0 \text{ or } \varepsilon \text{ or } C)$.

– If $X_j = mC + \varepsilon$, with $m = 0,1,\ldots,k$, then $X_{j+1} = X_j + (0 \text{ or } C)$.

The graph is simplified by eliminating the vertices associated with non-feasible values of stock levels, i.e. the X_j that are such that:

$$X_j - \sum_{i=u+1}^{j} d_i < 0 \qquad [4.95]$$

for $j = u+1, \ldots, v-1$ in the case without shortage:

$$X_j - \sum_{i=u-\alpha+1}^{j} d_i < 0 \qquad [4.96]$$

for $j = \alpha, \alpha+1, \ldots, v-1$ with shortage.

An initial vertex, written X^*, is added to the graph. Every path between X^* and X_v is a feasible production sequence. To solve $\overline{S}_{uv}$ and find $\overline{S}^*_{uv}$ and the associated cost d_{uv}, it is necessary to add a cost to each arc. We write this cost as $g(X_i, X_{i+1})$ defined as follows:

$$g(X_i, X_{i+1}) = p'_{i+1}(X_{i+1} - X_i) + h_{i+1}(I_{i+1}) \qquad [4.97]$$

for $i = *, u+1, \ldots, v-1$ and $X^* = 0$.

To solve a subproblem $\overline{S}_{uv}$, we must seek the shortest path in the graph thus defined. We summarize the presented method in Algorithm 4.1.

Algorithm 4.1 The Florian and Klein method

1: Step 1: Solution of the $\overline{S}_{uv}$
2: **for** $u = 0$ to $T - 1$ **do**
3: **for** $v = u + 1$ to T **do**
4: Test if $\overline{S}_{uv}$ is feasible, i.e. if $\sum_{i=u+1}^{v} d_i \leq (v - u) \times C$
5: **if** $\overline{S}_{uv}$ is feasible **then**
6: seek $\overline{S}_{uv}^*$: construction of the graph associated with the subproblem (the search for the states and the calculation of costs) and the search for the shortest path, of minimal cost d_{uv}
7: **end if**
8: **end for**
9: **end for**
10: Step 2 : Solution of the overall problem
11: $f = 0$
12: **for** $u = 0$ to $T - 1$ **do**
13: $f_u = min_{u<v\leq T}\{d_{uv} + f_v\}$
14: **end for**

4.5.4.2.4. Example of an application

Let us consider an example with $C = 7$, $T = 4$, $p_i'(x_i) = \delta(x_i) \times (6 - i) + 5x_i$, $h_i^-(I_i) = -2I_i$ and $h_i^+(I_i) = iI_i$. We consider the maximum number of periods of delay, α, to be unlimited. The demand for each period i, written as d_i, and the cumulative demand and the cumulative capacity are given in Table 4.11.

Period i	1	2	3	4
d_i	3	6	8	3
Cumulative demand	3	9	17	20
Cumulative capacity	7	14	21	28

Table 4.11. *An example of the Florian and Klein method*

Let us take, for example the subproblem $\overline{S}_{04}$. The cumulative demand is 20 and the total capacity available over these four periods is 28. The subproblem is therefore feasible. We must now determine the set of possible states for each period, that is, the set of cumulative production values for $i = 1, 2, 3, 4$. We have $k = 2$ and $\varepsilon = 6$. The subproblem consists of four periods; there is, therefore, one period in which the quantity produced is 6, two periods where the total capacity is produced and one period where nothing is produced.

The graph associated with this problem is given in the figure below. The cost of the shortest path between vertex X^* and each other vertex of the graph is shown beside each vertex. These values are calculated from the costs associated with each arc.

Period i	X_i
1	0;6;7
2	6;7;13;14
3	13;14;20
4	20

Table 4.12. *An example of the Florian and Klein method (1)*

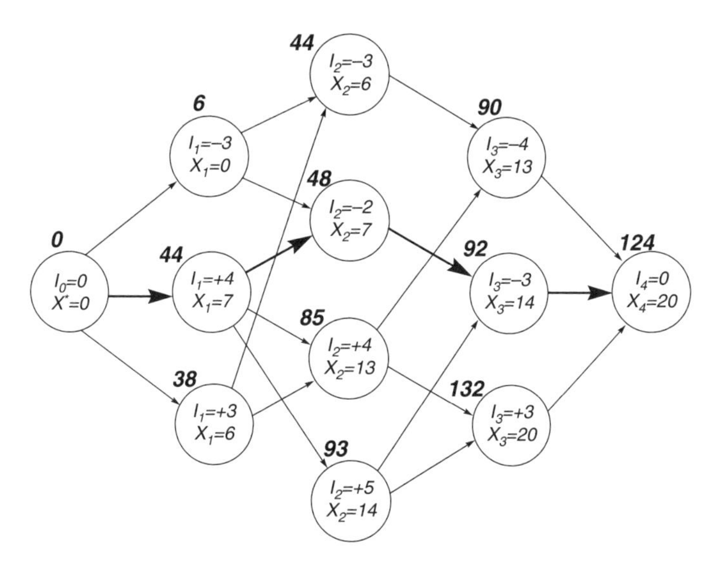

Figure 4.20. *An example of the Florian and Klein method (2)*

The shortest path is shown in bold on the graph. We find $d_{04} = 124$. Once the solutions to all of the subproblems have been found, we move onto the solution to the general problem, which is graphically shown in Figure 4.21.

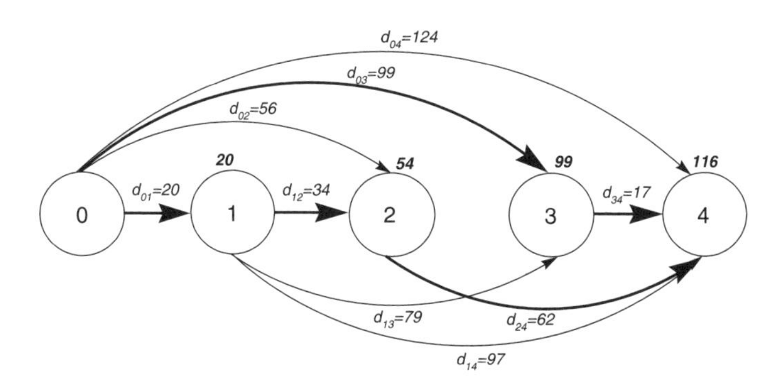

Figure 4.21. *An example of the Florian and Klein method (3)*

The minimum cost for covering the requirements over the four periods while respecting the constraint is 116. There are two paths that allow us to obtain this cost. They are shown in bold on the graph.

4.6. Quality management

The issue of quality lies at the center of business life. The classic epistemological approach to the subject is to separate quality into two parts: quality management and quality assurance. Others suggest distinguishing between tools and management/certification. None of these views are satisfactory.

4.6.1. *Evaluation, monitoring and improvement tools*

The tools necessary for the evaluation of the quality, the product or the performance of an organization are presented in this section. Without going into detail, we discuss the manner in which metrology enters into quality control. Next, we present a number of statistical tools necessary for the interpretation of measurements through various quality controls.

4.6.1.1. *The objective of metrology*

The implementation of measures of the quality of a process provides results that are used to make different types of decisions:

– acceptance or rejection of a product;

– setting a measuring instrument;

– setting a machine;

– validation of a process, etc.

The validity and relevance of this type of decision depend directly on the quality of the given information and, therefore, on the results of the measurements performed. We quantify the quality of a measurement by its uncertainty, thus allowing the estimation of its reliability.

4.6.1.2. *Concepts of error and uncertainty*

We will distinguish between the terms "error" and "uncertainty". Indeed, even after having evaluated the components of the error, there remains an uncertainty in the reliability of the result.

The goal is to report a result close to the true value. To achieve this objective, the operator responsible for the measurement will reduce the systematic errors by applying corrective measures. He or she will reduce random errors by repeating the measurement process.

Sources of error, or uncertainty, arise from diverse and varied factors. Overall, the different sources of error are due to five fundamental parameters identified under the title $5M$ (Figure 4.22). The $5M$s, shown in the Ishikawa diagram [ISH 90], are methods of measurement and work, means, materials, milieu (environment) and manpower.

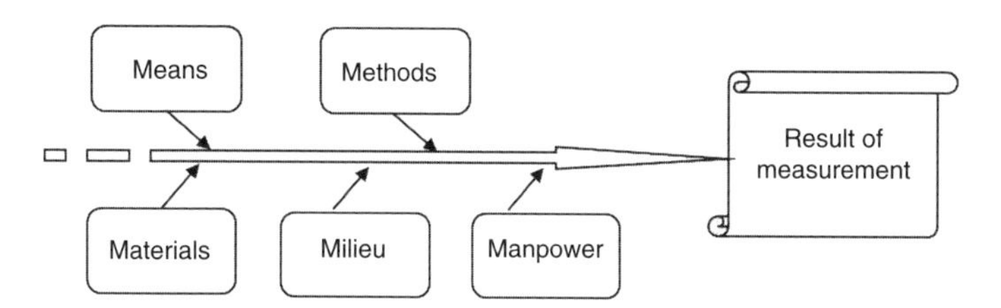

Figure 4.22. *The 5M principle (the Ishikawa diagram [ISH 90])*

The basic information is that the error, the uncertainty or any other unanticipated result is due to one of the five parameters stated above, or a combination of them.

The aim of quality control, regardless of the type, is to establish the offset on the one hand and to determine the source of non-quality on the other hand.

4.6.1.3. *Statistical quality control*

The ideal process for every industry would be to carry out an exhaustive control, a control of 100%, which consists of verifying all the pieces of their products to guard against anomalies. It is obvious that bearing the costs of such a draconian control is economically impossible.

It must also be noted that certain controls are destructive, and it would therefore be inappropriate to alter all of the pieces after having produced them. Nevertheless, there are non-destructive controls to be performed for 100% of the items when quality demands it. This is the case, for example, for items known as security parts.

4.6.1.4. *Stages of control*

The different stages of control are:

– Verification of the normality of the data. In effect, it is useless to analyze, review or recommend if the used data are incorrect.

– Verification of the capacity or capability of the production process or tool. This consists of verifying that the system possesses the technological, human and organizational competence to achieve a certain product. Capability consists of comparing two indicators – the demands of the product and the competence of the system.

– Implementation of a monitoring and control method during manufacture (control charts based on the characteristics to be studied).

4.6.1.5. *Tests of normality*

4.6.1.5.1. The normal distribution

The normal distribution governs the variations in numerous physical parameters. It is applicable when the dispersion of the variable is due to the influence of numerous mutually independent parameters whose effects are additive. Often, we assume that the distributions follow normal distributions. This is the case for the majority of distributions of continuous variables influenced by a large number of parameters of which none is dominant. However, we may encounter distributions that do not, generally, follow normal distributions. It is also possible that the information-gathering method itself leads to values following other distributions. Sometimes, it is the mixture of several populations of variables that modifies the distribution.

There are various methods to test the association with the normal distribution. Some are very elaborate, such as the chi-squared test, and others are graphical methods, such as frequency histograms or Henry's test.

A series of observations, or more precisely, samples, representing the evolution of a given phenomenon is the first step in every descriptive statistical study. In the case where the samples are taken according to a well-established protocol, or more precisely at regular intervals, we have a time series. This means that a study of the chronology of the variable is performed. Some techniques exist for determining the statistical nature of the variable's behavior, such as those presented below.

4.6.1.5.2. Histograms

The construction of a histogram from the series of values is often the simplest solution. The histogram of a normal population has the general shape of the normal distribution curve (Figure 4.23).

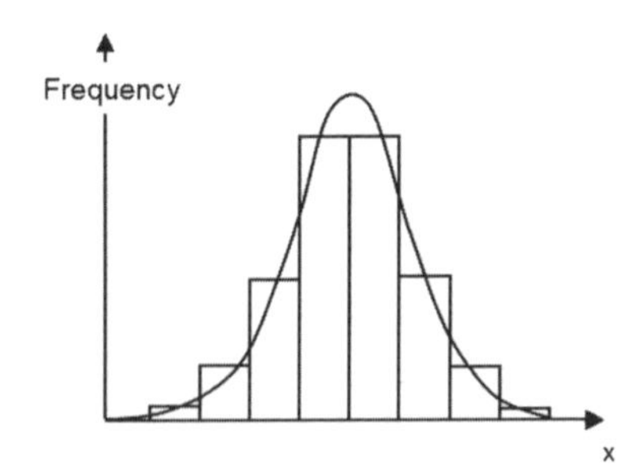

Figure 4.23. *A unimodal normal distribution*

It may be that the histogram is bimodal (Figure 4.24), i.e. there are two nodes. This generally reflects the mixture of two populations. This is the case for a series of items manufactured in parallel using two differently set means of production, for example. It may also be multimodal, when there is a mixture of n populations with different means.

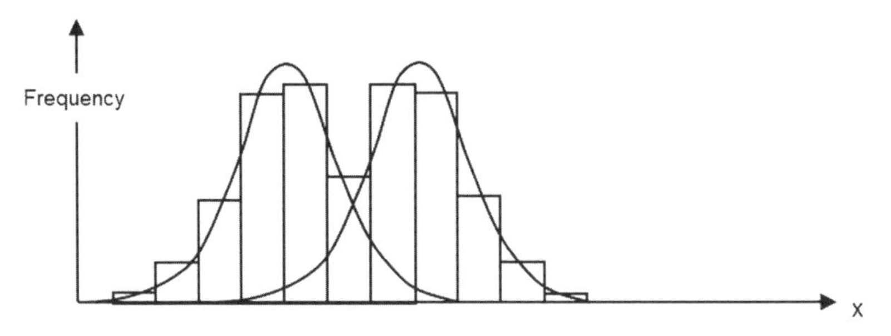

Figure 4.24. *A bimodal normal distribution*

4.6.1.5.3. Henry's test

Henry's test is a graphical test that allows us to verify the hypothesis of the variable's normality by examining the alignment of representative points in the distribution. The points in question are the different observations associated with their corresponding cumulative frequencies. The procedure consists of representing the cumulative frequencies of the considered samples in a Gauss arithmetic reference. This entails:

– The creation of equivalence classes of observations of a similar size.

– The creation of a cumulative frequency table corresponding to each class of sample. The mean X and the standard deviation (of all the frequencies) of the set of observations are calculated. Each observation is then considered as being the value of the cumulative distribution function of the variable in the case where it follows the normal distribution with mean X and standard deviation n (where n is the number of observations considered, or the size of the population).

– The representation of the class centers (the abscissa) as a function of the cumulative frequencies (the ordinate).

– The verification of the linear alignment of the points thus obtained.

The conclusion on the normality of the data is made on the basis of visual analysis. This method is very simple and very efficient. It is also possible to conclude this method without drawing the graph, but rather by calculating the correlation coefficient between the two data series. This allows a direct decision, assuming the level of the correlation coefficient reporting on the linear dependence between the two series is clearly defined (this decision value is generally approximate to 1).

EXAMPLE 4.18.– Let us consider a variable X with nine variation classes and a total of 100 values. The classes are of length 0.2, ranging from 209.1 to 210.9. Consider the variable Y that represents the distribution and frequency of the different values of X. The analysis of Y allows conclusions to be made about X. The frequencies are given in Table 4.13. We have $n = 100$.

Class	Lower limit	Upper limit	Occurrences	Center	Frequency (%)	Cumulative frequency (%)
1	209.1	209.3	1	209.2	1.00	1.00
2	209.3	209.5	7	209.4	7.00	8.00
3	209.5	209.7	17	209.6	17.00	25.00
4	209.7	209.9	23	209.8	23.00	48.00
5	209.9	210.1	22	210	22.00	70.00
6	210.1	210.3	20	210.2	20.00	90.00
7	210.3	210.5	9	210.4	9.00	99.00
8	210.5	210.7	0	210.6	0.00	99.00
9	210.7	210.9	1	210.8	1.00	100.00

Table 4.13. *Data for Henry's test*

The mean frequency equals 11.11%, $s^2 = 0.80\%$ and $s = 0.90\%$. The variable Y follows the normal distribution with mean 11.11% and standard deviation $\frac{s}{\sqrt{n}}$.

For the creation of a point in the Gauss arithmetic series, we proceed as follows:

– $Y = 209.2$, we seek: U_y with $P(U < U_y) = \text{Cumul.Freq} = 1\%$.

– $Y = 209.4$, we seek: U_y with $P(U < U_y) = \text{Cumul.Freq} = 8\%$.

All of the values of U_y are given in Table 4.14.

Abscissa Y	Cumulative frequency (%)	U_y
209.2	1.00	0.090
209.4	8.00	0.099
209.6	25.00	0.105
209.8	48.00	0.111
210	70.00	0.116
210.2	90.00	0.123
210.4	99.00	0.132
210.6	99.00	0.132
210.8	100.00	0.184

Table 4.14. *An example of Henry's test*

The study of this straight-line method allows us to conclude that the random variable Y follows a normal distribution with parameters $m = 11.11\%$ and $\sigma = 0.08\%$. This result can be obtained by calculating the correlation coefficient between Y and U_y, which is 0.9. This shows the strong linear dependence between the variables, and therefore the straight line does, indeed, exist.

Once we have established the distribution that the random variable Y follows, we obtain the distribution of X and its parameters by efficiently calculating the mean and the standard deviation.

4.6.1.5.4. Chi-squared test

This hypothesis test is used for both qualitative and quantitative characteristics (generally frequencies) arranged into classes. It is valid for large sample sizes, and requires at least 10 classes with 4–5 observations per class. This method involves certain parts already seen in the previous method. The implementation of the chi-squared test is independent of the distribution associated with the considered variable. The method proceeds following these steps:

– The creation of observations in equivalence classes of the same size.

– The calculation of the mean $\overline{x}$ and the standard deviation s of the data set.

– The calculation of $u_i = \frac{e_i - \overline{x}}{s}$, where e_i is the value of the upper limit of class i.

– The search for $F(u_i)$ in the table of the standard normal distribution with $\mathrm{F}(-\infty) = 0$ and $\mathrm{F}(+\infty) = 1$.

– The calculation of the theoretical frequencies:

- n'_i such that $n'_i = n[F(u_i) - F(u_i - 1)]$ and $n' = n[F(u_i)]$;

- if $n' > 5$, the value is correct; otherwise, we merge two classes to satisfy this constraint.

– The determination of the number of degrees of freedom (dof) (in general, this is equal to the number of retained classes minus 3): dof $=$ number of retained classes minus p minus 1, where p is the number of parameters in the studied distribution. If $p = 2$ as in the previous example in the case of the normal distribution, we obtain dof $=$ number of retained classes minus 3.

– The choice of the risk α of rejecting a true hypothesis (we generally take 5%).

– The calculation of χ^2:

$$\chi^2 = \sum_{i=1}^{\mathrm{numberofclasses}} \frac{(n'_i - n_i)^2}{n'_i} \qquad [4.98]$$

– The examination of the theoretical χ^2 corresponding to the accepted risk α and to the number of degrees of freedom.

– The comparison of the calculated χ^2 with the theoretical χ^2. If the calculated value is less than the theoretical value, the hypothesis about the nature of the distribution (the normality of the distribution) may be accepted, with a risk of error of α.

EXAMPLE 4.19.– Let Y be a variable whose values are grouped into the seven classes shown in Table 4.15, with $n = 50$, $X = 2,400$ and $s = 285.67$.

Class	Min	Max (e_i)	Number
1	1.700	1.900	3
2	1.900	2.100	6
3	2.100	2.300	5
4	2.300	2.500	18
5	2.500	2.700	12
6	2.700	2.900	4
7	2.900	3.100	2

Table 4.15. *The numerical values for the example of the chi-squared test*

The first phases of transformation are given in Table 4.16.

U_i	$F(U_i)$	$P_i = F(U_i) - F(U_{i-1})$
−1.75	0.04	0.04
−1.05	0.14	0.10
−0.35	0.36	0.21
0.35	0.63	0.27
1.05	0.85	0.21
1.75	0.95	0.10
2.45	0.99	0.03

Table 4.16. *The transformation phases for the chi-squared test*

The theoretical size of the classes is given in Table 4.17.

n_i'	Adjusted class	$(n_i - n_i')^2$	$(n_i - n_i')^2/(n_i')$
2.00	<5		
5.34	$2.00 + 5.34 = 7.34$	2.75	0.37
10.81	10.81	33.83	3.12
13.68	13.68	18.61	1.36
10.81	10.81	1.40	0.12
5.34	$5.34 + 1.64 = 6.98$	0.96	0.13
1.64	<5		

Table 4.17. *The theoretical size of the classes*

After calculation, we obtain $\chi^2_{\text{calculated}} = 5.13$, dof $= 5 - 3 = 2.00$, $\alpha = 0.05$, $\chi^2_{\text{theoretical}} = 5.99$. The normality of the data is accepted with the parameters calculated previously.

4.6.2. *Types of control*

Depending on the established criteria, we distinguish between several categories of quality control. The classification of controls may be made with respect to the controlled quantities (a sample or the totality), the position of the controlled product with respect to the treatment process (on entrance: reception, inside: manufacture or on exit: delivery) and/or by the verification protocol (visually, with measurements, on some attributes, etc.).

Here, the subject of control will be addressed by distinguishing, from a technical perspective, four pairs of decisions, which are single-decision control of measurable characteristics, single-decision control of product attributes, measurable-characteristic-monitoring control and attribute-monitoring control. Note that we mainly consider the configuration of physical production systems (automotive, textile, parts, items, etc.) with known limits.

Figure 4.25 shows, for a given company, the types of control that are addressed here. They are essentially of two types: reception control and manufacturing control. Recall that reception control coincides with output (delivery) control.

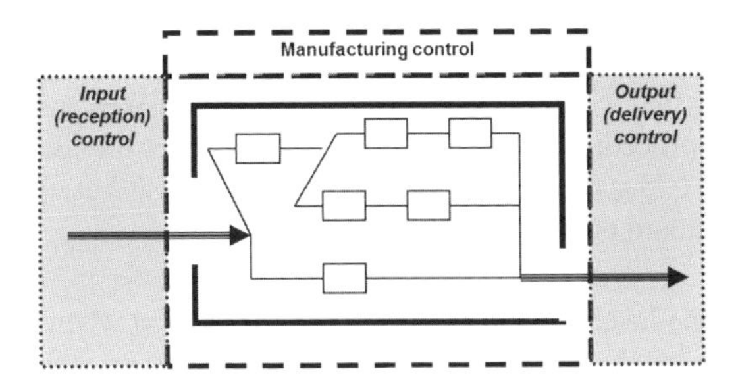

Figure 4.25. *Types of control*

4.6.2.1. *Reception or final control*

Reception control is an essential and decisive part of the quality policy adopted by each company, as it allows the inputs to be filtered and controlled. It is said to be single-decision because it is concerned with deciding whether or not a batch of (arriving) products is fit for service. This control has the main objectives of:

– deciding the acceptance or rejection of a batch;

– comparing the quality of batches coming from multiple suppliers;

– assessing the quality level of a series of batches to determine whether or not the control rules are strict enough.

On the one hand, this control allows rational decisions to be made, and on the other hand, it allows us to rate suppliers based on the quality of their production. To perform these controls, it is necessary to put in place a control plan, also called a control policy. The work of Amodeo and Yalaoui [AMO 05] contains a detailed explanation of the different aspects of reception control: applications, control plans, partnership agreements, the calculation of different parameters, different types of control, etc.

4.6.2.2. *Reception control by measurement*

Measurement-based reception controls are valid in the case where defects may be identified by measurement and are normally distributed. In the case where there is no proof of the normality of the distribution followed by the defects, it is necessary to carry out a suitability test, which amounts to performing a χ^2 test or a Henry's test. This type of control is more economic in terms of sample sizes compared to counting controls. However, checking each product requires more work, as making measurements takes much more work than counting its defects. Moreover, measurement-based control is subject to frequent errors due to the technicians performing the measurements, or in the interpretation of their results, which makes this type of control risky as the economic losses may be colossal.

Here, we may also construct simple, double, multiple and progressive plans as in control by counting that is addressed previously. We may also use normal, reduced and reinforced control procedures. However, we also encounter the use of predetermined plans. The NFX 06 023 standard provides some of these plans based either on a known or an unknown standard deviation.

The notion of the level of control enters into this context when the plan is chosen as a function of the batch size. The standard defines three control levels:

– Level I: the level of criteria that are particularly difficult to control.

– Level II: the standard level that is usually chosen.

– Level III: the level of criteria that are particularly easy to control.

In addition to these general-purpose levels, the norm provides two special-purpose levels, denoted S_3 and S_4, which are only adopted if compelling reasons oblige us to only take samples of small sizes. The controlled quantities are low but the efficiency is mediocre.

4.6.2.2.1. Plans based on the (known or unknown) standard deviation

We distinguish between the σ-method, which is used when the standard deviation is known, and the s-method, which is used when the standard deviation is unknown, where s is the estimation of σ calculated over each sample.

In both cases, we can construct:

– a control chart of the mean, where the acceptance limits defined by the plan are shown;

– a control chart of the standard deviation or the range, where the classical control limits are shown.

The first chart will let us accept or reject a batch according to the position of the mean relative to the acceptance limits. The second chart will let us, in the case where the standard deviation is known, verify the stability of the dispersion and, in the opposite case, use the s-method if the dispersion is stable for a certain period of time.

4.6.2.2.2. Control plans based on tolerance limits

Let us consider the following data: the acceptable quality limit (AQL), the rejectable quality limit (RQL), α and β, and the lower and upper defect tolerances (T_l and T_u). We assume a control policy with size n and a decision limit L (to obtain A and R).

From the tolerances T_u, we have:

$$L = T_u - |U_{AQL}|\sigma + |U_\alpha|\frac{\sigma}{\sqrt{n}} \quad [4.99]$$

$$L = T_u - |U_{RQL}|\sigma - |U_\beta|\frac{\sigma}{\sqrt{n}} \quad [4.100]$$

Hence

$$|U_{AQL}| - |U_{RQL}| = \frac{|U_\alpha| + |U_\beta|}{\sqrt{n}} \quad [4.101]$$

From this, we obtain the decision parameters:

$$n = \left[\frac{|U_\alpha| + |U_\beta|}{|U_{AQL}| - |U_{RQL}|}\right]^2 \quad [4.102]$$

$$L = T_u - \left[\frac{|U_\alpha||U_{AQL}| + |U_\beta||U_{AQL}|}{|U_\alpha| + |U_\beta|}\right]\sigma \quad [4.103]$$

The procedure of this control method then consists of taking n items, and writing m for the measurements of the characteristic. We can distinguish the two following cases:

– $m \leq L$: acceptance;

– $m > L$: rejection.

4.6.2.2.3. Simple sampling plan

In the most common cases, the batch size is fixed, and we choose an AQL value. For the usual level (II), the tables give a letter code. For this letter code, normalized tables give a plan with the constant allowing the calculation of the rejection or acceptance criteria, as well as the values of p_{95} and p_{10} corresponding to the supplier and customer risks.

Other ways of selecting a plan may be envisaged depending on the parameters that we wish to favor: AQL, batch size, sample size, p_{10}, p_{95}, average outgoing quality level (AOQL) or control level. Depending on the case, it will be necessary to use other tables, which are found in the Association Française de Normalisation (AFNOR) standard.

4.6.2.2.4. Progressive sampling plan

The NFX 06 023 and 06 025 standards propose a procedure for formulating measurement-based progressive control plans. The principle is the same as for the progressive counting control, but the implementation is difficult because it requires time and qualified personnel. The following concerns the case where the standard deviation is known and a single tolerance limit is considered. For two separate tolerance limits, we carry out the control for each limit. The batch will only be accepted if it is separately accepted for each limit. In the case of progressive control, a decision is made after the control of each individual. The plan is constructed taking two points on the efficiency curve. The decision rule, which comes from Wald's theory of sequential tests, is as follows: x_j being the measurement made in individual j, we write $y_j = x_j - T_l$ for a lower limit or $y_j = T_u - x_j$ for an upper limit.

The batch is accepted if:

$$\sum_{j=1}^{n} y_j \geq h_1 + s.n \qquad [4.104]$$

The batch is rejected if:

$$\sum_{j=1}^{n} y_j \leq -h_2 + s.n \qquad [4.105]$$

Sampling continues if:

$$-h_2 + s.n < \sum_{j=1}^{n} y_j < h_1 + s.n \qquad [4.106]$$

where

$$h_1 = \frac{\ln(\frac{1-\alpha}{\beta})}{k}.\sigma \qquad [4.107]$$

$$h_2 = \frac{\ln(\frac{1-\beta}{\alpha})}{k}.\sigma \qquad [4.108]$$

$$k = u_{1-p_1} - u_{1-p_2} \qquad [4.109]$$

$$s = \frac{\sigma}{2}.(u_{1-p_1} + u_{1-p_2}) \qquad [4.110]$$

u_{1-p} being the quantile of order $1 - p$ of the standardized normal variable.

This decision rule may be represented graphically as in the case of progressive counting control. To avoid being held in the indecision zone indefinitely, we stop the control by truncation when we have, for example, taken 1.5 times the number of individuals given in a simple plan. The decision is made by considering the closest straight line.

4.6.2.3. *Manufacturing control*

This type of control (Figure 4.26) consists of the monitoring of measurable characteristics. The objective of this quality control is the detection of non-compliant parts at the time and place at which they are produced, or the monitoring of the possible maladjustments of the manufacturing process in order to quickly undertake corrective action. This type of control is included in the family of sampling-based control. From the different readings, we may put in place an improvement process that relies on:

– the analysis of results;

– the calculation of new characteristics;

– the setting of tighter limits.

From this approach, we again arrive at two main objectives of this type of control: the prevention of the occurrence of deviations to minimize or even eliminate the manufacture of products that do not comply with specifications and the improvement of reproducibility by managing the variability of the process.

Manufacturing control consists of performing a statistical test repeated over successive samples taken. At each sampling, we check if the result (a characteristic to be studied, such as the mean, the standard deviation, the range, etc.) lies in the confidence interval. This interval is represented by limits drawn on a graph or the sheet of statements. The study and monitoring of manufacturing is included in the rationale of supervision and support.

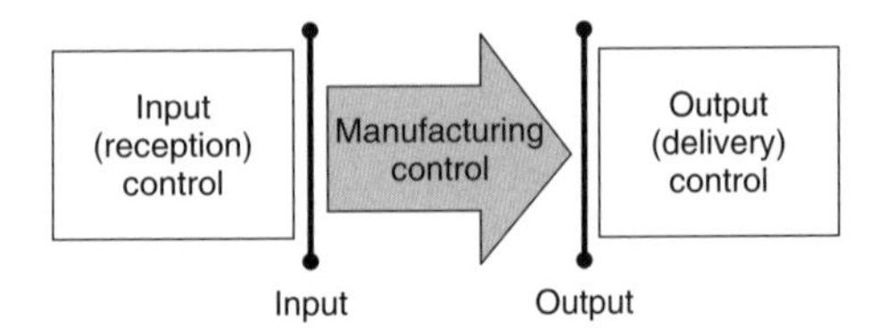

Figure 4.26. *Manufacturing control*

The main questions to which this type of control attempts to respond are the following. Does the system actually provide the required product? Does it follow all the specifications and demands of the originator of the order (internal customer, external customer, etc.)? Is there truly a stability within the type of product realized?

4.6.2.3.1. Capability

The concept of capability is a widespread notion in production workshops. Production managers always have a vague idea of the quality of the machines available in the workshop, and for that reason they are generally incapable of giving a numerical value behind this impression. Yet it is fundamental, when talking about quality, to know exactly what the process is capable of compared to what is asked of it. To be able to correctly realize a product, we must check the capacity (the capability) of the process and its components. Generally, every process seeks to produce a target reference according to a given protocol. However, every process is equally subject to the influence of different families of parameters contained in the 5 Ms. The capability will consist of quantifying the performance of the process compared to the known demands. It may be defined as a quantitative interpretation of a vague notion, and consists of estimating the ratio of the required tolerances and the actual dispersion of the method around a target value. The tolerances are represented by the precision interval accepted for a product around its target (Figure 4.27). We define the tolerance interval by $TI = T_u - T_l$, where T_l and T_u are the lower and upper tolerances, respectively.

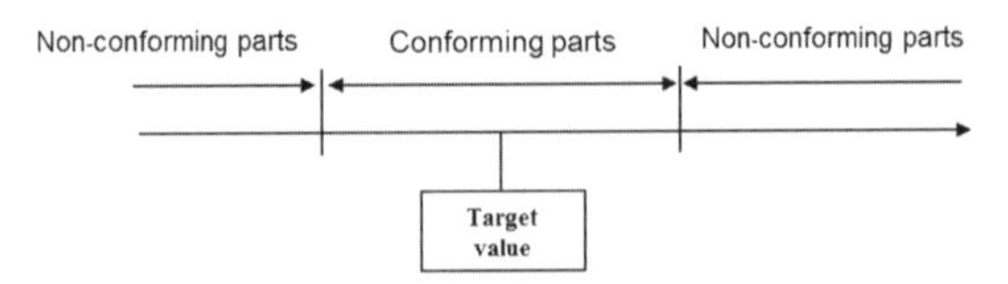

Figure 4.27. *The tolerance interval*

The evaluation of the real dispersion of the process or system consists of quantifying, as a function of the work time frame and the factors coming into play, the standard deviation of the population representing a type of production. The dispersion specific to each change is represented by an interval $I_p = 6 \times \sigma$. The capability is then as follows:

$$C = \frac{TI}{I_p} = \frac{TI}{6\sigma} = \frac{T_u - T_l}{6\sigma} \qquad [4.111]$$

It should be noted that there are two types of capability coefficient:

– The first only shows the dispersion:

- C_m: machine capability;

- C_p: process capability.

– The second shows the centering and the dispersion:

- C_{mk}: machine capability;

- C_{pk}: process capability.

4.6.2.3.2. Machine capability

This indicator is identified on the basis of the dispersion of the system related solely to machines. In other words, it only concerns the variation caused and generated by machines, or more generally, the production tools. This also means that in the 5M family, the four other Ms are assumed not to influence the system, or to be stable. The machine capability therefore represents the measurement of the performance of the machine only, independent of other factors. This leads us to consider that the dispersion becomes $C_m = \frac{T_u - T_l}{6\sigma}$, where m is the machine (tool) dispersion. We distinguish between two types of machine capability: C_m and C_{mk}.

4.6.2.3.3. The C_m machine capability

For this type of capability, the following facts are taken into account:

– An instantaneous, point dispersion due to the tool or machine: σ_m, which is also denoted by σ_i.

– The study of the capability of a process is preceded by the study of the capability of the tool (machine). However, this is not a conclusion, because the time frames, the objectives, the demands and even sometimes the customers are not the same in the cases of very short, short, medium or long terms.

– This dispersion is mainly due to the machine. Sampling is done over a relatively short period of time and in stable conditions, so as to ensure the stability of other parameters ($4M$s).

– We use a limited number of parts (observations) manufactured by the same machine (tool). The determination depends strongly on operating times.

The results are expressed in the form of a distribution of values, the most often being the normal distribution with a position parameter (the mean) and a dispersion parameter (the standard deviation).

Note that:

$$C_m = \frac{TI}{6.\sigma} \quad [4.112]$$

To not have a rejection, the condition $C_m \geq 1.33$ must be satisfied. This threshold is established with:

$$C_m = \frac{8\sigma_m}{6.\sigma_m} \quad [4.113]$$

We have the following different cases, enabling a conclusion to be made after the calculation of C_m:

– $C_m < 1$: the machine is not capable.

– $C_m = 1$: the machine is capable.

– $C_m = 1.33$: the machine is correct.

– $C_m = 1.5$: the machine performs well.

The capability index of the machine is calculated from data recorded in a very short period of time, including only the dispersion of the machine, outside of any modifications in other production factors (materials, manpower, etc.). If the calculation of $T_u - T_l$ is done without difficulty, it is necessary to define the method allowing the calculation of σ_i (the instantaneous standard deviation). The calculation is performed using at least 50 consecutive parts, trying to best neutralize external influences such as changes of operator and modification of the work environment (temperature, etc.).

4.6.2.3.4. The C_{mk} machine capability

We have studied the machine capability C_m that essentially includes the dispersion. The implementation of an indicator allowing the quantification of the maladjustment of the limits with respect to the target is important. This consideration gives more pertinent indicators, but we must, before checking the alignment, ensure the overall performance. C_{mk} takes into account, by construction, the alignment relative to the target.

This type of capability is given as follows:

$$C_{mk} = \min\left\{\frac{T_u - \overline{x}}{3\sigma_i}; \frac{\overline{x} - T_l}{3\sigma_i}\right\} \quad [4.114]$$

where $\overline{x}$ is the mean of the values observed.

4.6.2.3.5. The process capability

In addition to the machine capability, the process capability C_p gives an evaluation of the entire process. This capability may be considered as an overall, long-term indicator of the process, in contrast to the machine capability C_m. The capability C_p is broader as it is calculated using not an instantaneous point dispersion but the total dispersion of the whole process. The dispersion representing the system is generated by the different factors of the 5M system. Like C_m, C_p comes in two variants: C_p and C_{pk}.

The process capability C_p is based on the following data:

– The overall dispersion of the entire process, written as σ_g.

– This capability gives an idea of the general performances of the process, and also over longer terms.

– This dispersion is due the the $5Ms$. Sampling is done over a longer or shorter period of time depending on operational times, enabling the involvement of the 5 Ms. Therefore, this requires some weeks, or sometimes a few months.

Note that $C_p = \frac{TI}{6\sigma_g}$ and we take exactly the same limits and make the same comments as for C_m.

Regarding the C_{pk} process capability, as for the relation between C_m and C_{mk}, C_p includes only the dispersion, while C_{pk} takes account of the centering with respect to the target. The rationale and utility of both indicators are presented in previous sections. C_{pk} is a reference indicator:

$$C_{pk} = \min\left\{\frac{T_u - \overline{x}}{3\sigma_g}; \frac{\overline{x} - T_l}{3\sigma_g}\right\} \quad [4.115]$$

where $\overline{x}$ is the mean of the observed values.

4.6.2.3.6. The target capability: C_c (or C_{pm})

This indicator, known as the target capacity, is based on the study of the loss function. The basic principle consists of studying the performance of the process by seeking to best determine them. We have already considered the dispersion of the system with respect to reference periods, and we will now outline the dispersion with respect to objectives. From the same manufacturing control, the purpose of this monitoring is to visualize the production relative to target characteristics and their tolerated limits. Hence, an indicator that would give the process capability while fully integrating this target production is of great importance. The procedure to be implemented in order to achieve this objective must be developed and strictly followed. Across multiple studies performed, this type of target indicator relates, in a specific way, to the two aforementioned types of dispersion (σ_i and σ_g). There are numerous studies in the literature that provide a better understanding of this function and the analysis of processes [LYO 97].

In summary, we have:

– The overall dispersion of the process: σ_c.

– The consideration of the dispersion and the centering.

– The indicator C_c is based on the Taguchi loss function (L). The mean loss L for a batch of mean $\overline{x}$ and standard deviation σ is $L = (\sigma^2 + (\overline{x} - \text{Target})^2)$, where K is a very large number.

– The consideration of the loss of system performance for the outcome with respect to a target value.

The indicator C_c reflects the loss (in the Taguchi sense) due to the maladjustment of the process. We define C_{pm} by the following relation:

$$C_c = \frac{TI}{6\sigma_c} = \frac{TI}{6\sqrt{\sigma_g^2 + (\overline{x} + \text{target})^2}} = \frac{TI}{\sqrt{1 + 9 \times (C_p - C_{pk})^2}} \qquad [4.116]$$

As a function of different values and the centering of the process, we find the conclusions provided by the combination of the indicators C_p and C_{pk}. Among the possible configurations of C_c, we note that if $C_c = C_p$, the process is perfectly set, and if C_c decreases, the maladjustment increases.

4.6.2.4. *Control charts*

We find different definitions of the notion of a control chart in the literature [BAI 80, LAM 89, LYO 97]:

– A control chart consists of a drawing of a statistical measure as a function of time, evaluated from a set of samples or subgroups, and on which we indicate the control limits.

– A control chart is a graph that represents successive images of the production, taken at regular intervals (the sample frequency), from parts taken from production (samples) and on which we perform calculations (mean, dispersion, etc.) that are reported on the graph(s) of the chart.

4.6.2.4.1. Description of a control chart

A control chart (Figure 4.28) usually comprises three main lines: the two outer lines indicate the upper control limit (CL_U) and the lower control limit (CL_L), whereas the central line usually represents the general mean of the statistical measure evaluated on the set of samples taken [BAI 80].

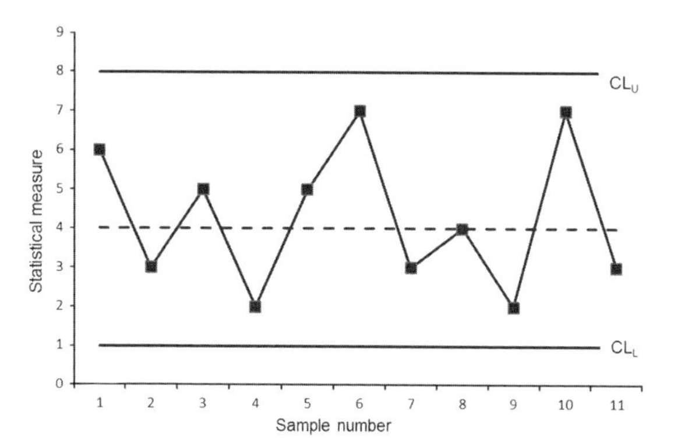

Figure 4.28. *A typical control chart*

Control charts allow the variability of the process to be visualized and followed by distinctly identifying the modes of variation. These may or may not be capable of being statistically modeled (by associating a random distribution such as a normal distribution and a binomial distribution). The possibilities of modeling perturbations are possible due to the precise identification of the origin of the variation. These are called assignable causes [BAI 80, DUR 98], and non-assignable causes in the opposite case.

We may distinguish between two large families, as for reception control, namely:

– Control charts for measurable quantities (measurement-based quality control), including those for the central tendency, the dispersion, etc.

– Control charts for non-measurable quantities or qualifiable characteristics (attribute-based quality control) such as number, proportion and binomial or Poisson distributions.

In the first case, the control is performed on characteristics that may be measured, such as diameter, weight and density, whereas in the second case, the result of the control related to the characteristics will be an attribute (conformity or non-conformity). The steps of a manufacturing control are as follows:

– Choose the characteristics to be monitored.

– Understand and master the methods of recording and analyzing the collected data.

– Study preliminary manufacturing performance indicators. Using the capabilities, the procedure is simple:

- If the indicators show insufficient performances, then we must intervene, setting, modifying and making the necessary repairs before the start of production.

- If the indicators are good, depending on the adopted limits, then launch the manufacturing system and efficiently monitor it. Monitoring is more efficient if the measured parameters are the right ones and are correctly sampled at appropriate times, or precisely, we can say, if the protocol is pertinent.

4.6.2.4.2. Sampling protocol

A minimum number of measurements are always necessary in order to ensure the significance of the data before any possible interpretation. Defining the protocol consists of establishing the number of parts (items) to sample for each observation, and giving the outcome frequencies of these samples. There are numerous sampling techniques available, depending on the parameters under consideration, the aim of the control, etc. Among the multitude of existing techniques, two are presented here. They differ according to the parameters taken into consideration, but provide the basic data for every protocol.

Method A: this is a basic method. It begins with the normalized value for the size of the observations and also for their outcome frequencies. The considered values are as follows:

– Every sample or observation does not exceed six values (parts, items) with a minimum of four values. Thus, we have a sample size n such that $4 \leq n \leq 6$.

– Depending on the rhythm of production, the observations must represent between 5% and 10% of the reference value. This translates into the checking of 5–10% of the number of items manufactured per hour. From these values and sample sizes, the sampling frequency is established.

Method B: this method, in contrast to the previous method, integrates many more parameters of the manufacturing system. It does not simply ensure the representativeness of the samples but inserts them into the specific configuration of the system through the production rhythm, the maintenance mode and the feasibility of the observations. The parameters integrated by this method are:

– The rhythm, or more precisely, the production rate of the tool (machine) per unit of time (hour): C.

– The monitoring and maintenance modes of the production system are taken into account. The justification is evident, and it comes from the fact that every manipulation of the tool generates perturbations and modifications, and therefore changes its random behavior. As a result, the modeling of the causes of variations is changed, and may introduce interpretation errors. There are numerous indicators to translate these interventions, of which the parameter N_{IR} represents the number of parts manufactured between two adjustments.

– This method begins with a certain maximum and minimum sample size n. The values of n are the same as for method A ($4 \leq n \leq 6$).

– From n parts to be taken, we establish the batch concerned by this sample, which is $\sqrt{n \times N_{IR}}$ parts. This represents, with respect to the production, quantities over intervals of $\sqrt{n \times N_{IR}} \times \frac{60}{C}$ m.

EXAMPLE 4.20.– Let us consider a company E with a manufacturing rate of $C = 300$ parts per hour. The objective is to control a minimum of 5% of the parts produced. On the other hand, between two successive adjustments, the system produces $N_{IR} = 600$ parts. The ideal way is to take samples of size $n = 5$. Depending on the method, the control protocols are as follows:

For method A: 15 parts must be taken per hour. Thus, according to this method, samples of 20 parts must be taken every 20 m.

For method B: we consider $n = 5$ parts, and these lots are taken from $\sqrt{n \times N_{IR}} = 55$ parts. These lots represent, in relation to production, sequences of $\sqrt{n \times N_{IR}} \times \frac{60}{C}$ = 10.95. Thus, the protocol according to method B will consist of taking five parts every 10 m and 57 s, i.e. virtually every 11 m.

We readily see that when integrating more information, the treatment and monitoring methods differ (being stricter, and even heavier, as they are more frequent).

EXAMPLE 4.21.– The elaboration of the minimum quantity to be taken may be obtained by taking into account different parameters, including the risks α and β, the machine capability C_m and a proportion p of tolerated parts exceeding the limit.

From Figure 4.29, we note that AD = AB + BC + CD, i.e.:

$$\frac{TI}{2} = \left[U_\alpha \times \frac{\sigma}{\sqrt{n}}\right] + \left[U_\beta \times \frac{\sigma}{\sqrt{n}}\right] + [U_p \times \sigma] \qquad [4.117]$$

where:

– TI: the tolerance interval.

– U_α: the reduced risk variable associated with the risk of the first kind (of detecting a maladjustment that does not exist).

– U_β: the reduced variable associated with the risk of the second kind (of not detecting a maladjustment).

– Up: the reduced variable associated with the percentage of acceptable parts exceeding the tolerance.

– n: the minimum number of measurements in the sample.

– σ: the standard deviation of the total population.

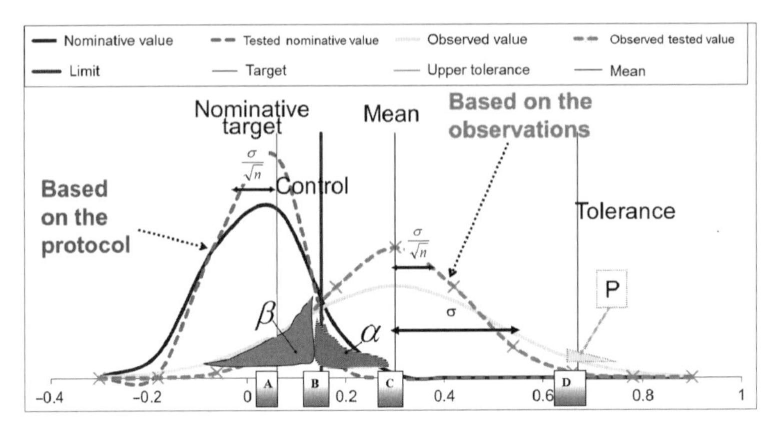

Figure 4.29. *Optimal sample size*

Then, $3C_m = \frac{U_\alpha}{\sqrt{n}} + \frac{U_\beta}{\sqrt{n}} + U_p$, given that $C_m = \frac{TI}{6\sigma}$.

Therefore, we obtain:

$$n = \left[\frac{U_\alpha + U_\beta}{3C_m - U_p}\right]^2 \quad [4.118]$$

This is the most common result. Following certain considerations and practices, the value of the sample size is:

$$n = \left[\frac{3 + U_\beta}{3C_m - U_p}\right]^2 \quad [4.119]$$

where $U_\alpha = 3$.

4.6.2.4.3. Measurement control charts

Once the sampling protocol is arranged and the indicators of the system are calculated, there remains to create the charts. Among the variety of existing charts and processes, the most frequently studied are:

– mean and range charts: $(\overline{X}, R)$;

– mean and standard deviation charts: $(\overline{X}, s)$;

– median and range charts: (M_e, R);

– individual observations and moving range charts: (X, R_m).

Therefore, we have a chart that reproduces the values of the characteristic with the mean, the median or even the individual observations, as well as a parameter representing the dispersion. This dispersion is the interpretation of the quality of the samples and observations, as the homogeneity of the observations is important, particularly if only the mean is reproduced on the monitoring chart. The observations necessarily concern several items. As for every chart, the stages of life are as follows:

– The elaboration and calculation of the control and monitoring limits:

- for the chosen characteristic;

- according to the adopted dispersion operator: standard deviation, range, etc.

– The physical production of the charts and the different tests.

– The use of the charts following the adopted protocol. Recording observations on the graph (the chart).

– Analysis and conclusion. This is translated, for example, by the study of points exceeding control limits, the detection of anomalies, policy proposals, etc.

4.6.2.4.4. Estimation of the dispersion

The estimation of the dispersion has an important impact on control as it will determine, in large part, the interpretation and analysis of the data observed. We then have two cases: the studied system is mastered and its dispersion σ^* is known, or it is necessary to estimate it in the most accurate possible manner. This estimation must be technically and practically efficient as this calculation is done for the first time to establish the charts and it is reproduced each time an observation is made. Thus, notions of complexity and calculation speed play a dominant role. This justifies the existence of two possible estimations of the dispersion: the standard deviation and the range.

The procedure consists of carrying out a first series of samples with a number of data that must be representative. The method for doing this must follow a well-established protocol, which will be the same for the rest of the manufacturing control system. These data, with K observations and n values each time, are identified and reported in Table 4.18.

For each observation $i = 1, \ldots, K$, we calculate the range R_i and the standard deviation σ_i as:

$$R_i = \text{Max}_j^n(x_{ij}) - \text{Min}_j^n(x_{ij}) \quad [4.120]$$

$$\sigma_i = \frac{1}{n}\sum_{j=1}^{n}(x_{ij} - \overline{x}_j)^2 \quad [4.121]$$

Sample i	x_1	x_2	...	x_j	...	x_n	σ_i	R_i
1								
2								
...								
i				x_{ij}				
...								
K								

Table 4.18. *The data of k observations and n values*

From these observations, we may determine the best estimation of the dispersion σ^* by following two different approaches:

$$\overline{\sigma} = \frac{1}{K}\sum_{i}^{K}\sigma_i \quad [4.122]$$

$$\overline{R} = \frac{1}{K}\sum_{i}^{K}R_i \quad [4.123]$$

Hence:

$$\sigma* = \frac{\overline{\sigma}}{b_n} = \frac{\overline{R}}{d_n} \quad [4.124]$$

The values of parameters b_n and d_n are given in Table 4.19 as a function of the sample size n.

We have then made an estimation of the dispersion $\sigma*$. There are other possible estimators, such as, s:

$$s = \frac{1}{(n \times K) - 1}\sum_{i=1}^{K}\sum_{j=1}^{n}(x_{ij} - \overline{x})^2 \quad [4.125]$$

where $\overline{x}$ is the mean of the observations.

4.6.2.4.5. Mean charts

Once the choice of characteristic is made and the method of evaluating the dispersion is chosen, there remains to create the charts. We have already noted the target value of the studied manufacture, followed by the mean value. This value is the most efficient empirical estimator of the expected value of the random phenomenon, that is the production under consideration. A $\overline{X}$ chart, or mean (drift) chart, is a graph on which the mean values of successive samples from the manufacturing process are shown. These values are connected to each other, preserving the chronological order of sampling. Note that to establish the total mean of the different observations $\overline{X}$, we must use the mean of each observation i, denoted by X_i. Therefore, we have:

$$X_i = \frac{1}{n} \sum_{j=1}^{n} X_{ij} \Rightarrow \overline{X} = \frac{1}{K} \sum_{i=1}^{K} X_i \qquad [4.126]$$

Size of each sample	d_n	b_n
2	1.128	0.564
3	1.693	0.724
4	2.059	0.798
5	2.326	0.841
6	2.534	0.869
7	2.704	0.888
8	2.847	0.869
9	2.970	0.888
10	3.078	0.903
11	3.173	0.930
12	3.258	0.936
13		0.941
14		0.945
15		0.945
16		0.952
17		0.955
18		0.958
19		0.960
20		0.962
21		0.964
22		0.966
23		0.967
24		0.968
25		0.970
26		0.971
27		0.972
28		0.973
29		0.974
30		0.975

Table 4.19. *Estimators of σ: b_n and d_n*

Manufacturing control depends, as stated previously, on the capability value of the system.

It is therefore necessary to distinguish between the cases where tolerances are and are not taken into account as well as the cases with and without an estimation of the dispersion.

4.6.2.4.6. Low capability or non-performance

If we return to the section on capability (section 4.6.2.3.1), this is deemed high and represents an effective system in the case where the capability is greater than 1.5 (for example $C_p \geq 1.5$). Conversely, in the case where $Cp < 1.5$, it is said to be low and corresponds to systems that are barely capable, or completely incapable. A control chart (Figure 4.30) is composed of a minimum of four limits, called upper and lower monitoring limits (ML_U(X) and ML_L(X)) and upper and lower control limits (CL_U(X) and CL_L(X)).

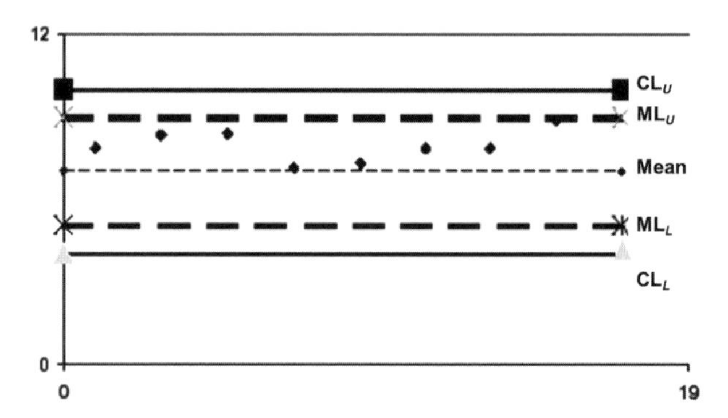

Figure 4.30. *An X control chart*

Independently of the type of known or estimated dispersion, supposing it is given by $\sigma*$, the different limits are:

– the upper control limit: $\mathrm{CL}_U(\overline{X}) = \overline{X} + 3\frac{\sigma*}{\sqrt{n}}$;

– the lower control limit: $\mathrm{CL}_L(\overline{X}) = \overline{X} - 3\frac{\sigma*}{\sqrt{n}}$;

– the upper monitoring limit: $\mathrm{ML}_U(\overline{X}) = \overline{X} + 2\frac{\sigma*}{\sqrt{n}}$;

– the lower monitoring limit: $\mathrm{ML}_L(\overline{X}) = \overline{X} - 2\frac{\sigma*}{\sqrt{n}}$.

In the case where $\sigma*$ is estimated by $\overline{\sigma}$ or R, we simply replace the corresponding values in the limits above.

4.6.2.4.7. High capability

In contrast to the previous case, for a high capability, with values of $C_p \geq 1.5$, it is imperative to include the tolerances in the establishment of the limits of the mean chart. This is justified by the fact that the system is highly capable, and we must not reject or refuse situations accepted by the customer, i.e. we must not be more strict than the demands imposed for the manufacture of a certain target characteristic. The limits are then given by the following formulas:

– the upper control limit: $\mathrm{CL}_U(\overline{X}) = T_u - 3\sigma * + \frac{\sigma *}{\sqrt{n}}$;

– the lower control limit: $\mathrm{CL}_L(\overline{X}) = T_l + 3\sigma * - \frac{\sigma *}{\sqrt{n}}$;

– the upper monitoring limit: $\mathrm{ML}_U(\overline{X}) = \overline{X} + 2\frac{\sigma *}{\sqrt{n}}$;

– the lower monitoring limit: $\mathrm{ML}_L(\overline{X}) = \overline{X} - 2\frac{\sigma *}{\sqrt{n}}$.

In this case, the monitoring limits do not change, and are not concerned in the first instance with the tolerances. This is due to the fact that monitoring may alert us, but by no means make removal or production stoppage decisions except in special cases. The fundamental change consists of taking a new estimation of the target value that is no longer made by a single value $\overline{X}$, but rather on the larger interval including, among others, $\overline{X}$, which is $[Ti + 3\sigma *, Ts - 3\sigma *]$.

4.6.2.4.8. Standard deviation (σ) charts and range (R) charts

A standard deviation σ chart is a graph on which the values of the standard deviation of successive samples taken from the manufacturing process are shown. These values are linked to each other, preserving the chronological order of sampling:

– the upper control limit: $\mathrm{CL}_U(\sigma) = \sqrt{\frac{\chi^2(0.999;n-1)}{n}}\sigma\chi * = B_{cs}.\sigma *$;

– the lower control limit: $\mathrm{CL}_L(\sigma) = \sqrt{\frac{\chi^2(0.001;n-1)}{n}}\sigma\chi * = B_{ci}.\sigma *$;

– the upper monitoring limit: $\mathrm{ML}_U(\sigma) = \sqrt{\frac{\chi^2(0.975;n-1)}{n}}\sigma\chi * = B_{ss}.\sigma *$;

– the lower monitoring limit: $\mathrm{ML}_L(\sigma) = \sqrt{\frac{\chi^2(0.025;n-1)}{n}}\sigma\chi * = B_{si}.\sigma *$

The values of parameters B_{cs}, B_{ci}, B_{ss} and B_{si} are given in Table 4.20 as a function of the sample size n.

The range (R) chart shows the values of the sample ranges.

These values are linked to each other, preserving the chronological order of sampling:

– the upper control limit: $\mathrm{CL}_U(R) = D_{cs}.\sigma *$;

– the lower control limit: $\mathrm{CL}_L(R) = D_{ci}.\sigma *$;

– the upper monitoring limit: $\mathrm{ML}_U(R) = D_{ss}.\sigma *$;

– the lower monitoring limit: $\mathrm{ML}_L(R) = D_{si}.\sigma *$.

Size of each sample	B_{ci}	B_{cs}	B_{si}	B_{ss}
2	0.001	2.327	0.022	1.585
3	0.026	2.146	0.130	1.568
4	0.078	2.017	0.232	1.529
5	0.135	1.922	0.311	1.493
6	0.187	1.849	0.372	1.462
7	0.233	1.791	0.420	1.437
8	0.274	1.744	0.459	1.415
9	0.309	1.704	0.492	1.396
10	0.339	1.670	0.520	1.379
11	0.367	1.640	0.543	1.365
12	0.391	1.614	0.564	1.352
13	0.413	1.591	0.582	1.340
14	0.432	1.570	0.598	1.329
15	0.450	1.552	0.613	1.320
16	0.467	1.535	0.626	1.311
17	0.482	1.520	0.637	1.303
18	0.495	1.505	0.648	1.295
19	0.508	1.492	0.658	1.288
20	0.520	1.480	0.667	1.282
21	0.531	1.469	0.676	1.276
22	0.541	1.458	0.684	1.270
23	0.551	1.449	0.691	1.265
24	0.560	1.439	0.698	1.260
25	0.569	1.431	0.704	1.255
26	0.577	1.423	0.710	1.250
27	0.584	1.415	0.716	1.246
28	0.592	1.408	0.721	1.242
29	0.599	1.401	0.727	1.238
30	0.605	1.394	0.737	1.235

Table 4.20. *Coefficients of the control and monitoring limits: standard deviation σ charts*

The values of parameters D_{cs}, D_{ci}, D_{ss} and D_{si} are given in Table 4.21 as a function of the sample size n.

4.6.2.4.9. Summary of charts

The main steps to be followed in the development of $\overline{X}$, R and σ charts are the following:

Step 1: sampling methods: size of the sample to be taken, control frequency, number of samples to be taken, data recording.

Step 2: calculation of means and ranges: calculation for each sample of the mean $\overline{X}$, the range R and the standard deviation σ.

Step 3: calculation of the control limits: calculation of the overall mean, the average range and the average standard deviation of the set of recorded data. Calculation of the upper and lower control limits for the three charts.

Step 4: identification of the control limits on the $\overline{X}$, R and σ charts: solid horizontal lines for the overall mean, the average range and the average standard deviation. Horizontal dotted lines for the upper and lower control limits of each chart.

Step 5: graphical representation of the means, ranges and standard deviations: plotting of the means on the $\overline{X}$ chart, indicating on the ordinates the means, and on the abscissas the identification of the sample (number, etc.). Connect the points. We do the same for the range and the standard deviation.

Step 6: diagnostics: analysis of the points on each control chart to determine the points exceeding controls and particular behaviors.

Size of each sample	D_{ci}	D_{cs}	D_{si}	D_{ss}
2	0.00	4.65	0.04	3.17
3	0.06	5.06	0.30	3.68
4	0.20	5.31	0.59	3.98
5	0.37	5.48	0.85	4.20
6	0.54	5.62	1.06	4.36
7	0.69	5.73	1.25	4.49
8	0.83	5.82	1.41	4.61
9	0.96	5.90	1.55	4.70
10	1.08	5.97	1.67	4.79
11	1.20	6.04	1.78	4.86
12	1.30	6.09	1.88	4.92

Table 4.21. *Coefficients of the control and monitoring limits: range (R) charts*

4.6.2.4.10. Attribute-based control charts

As mentioned previously, there are different types of charts, namely measurement-based and attribute-based. The approach and process of the latter charts are similar to that of the former. The fundamental notions of this type of chart are related to the decision-making mode. The decision consists of judging the conformity or non-conformity of a process. It is also necessary to distinguish between non-conforming processes, to classify products either outside of technical specifications or presenting important (serious) defects necessitating their rejection, and non-conformity referring to defects that do not necessarily lead to rejection, and that are taken into account in the measurement of quality. A product is judged as non-conforming in the case where it meets a non-conformity criterion on a pre-established list.

As for every control policy, a series of parameters are to be found, namely n_i, the number of items controlled in a sample or subgroup; np_i (also written d_i), the number of non-conforming items in observation i; p_i, the proportion of non-conforming items in sample i; and K, the total number of control samples. There are numerous charts said to be attribute-based, including the following:

– a chart of the population of non-conforming items from samples that are not necessarily of the same size: a (P) chart;

– a chart of the number of non-conforming units from samples of constant sizes: an (*NP*) chart;

– a chart of the number of non-conformities from samples of constant sizes: a (C) chart;

– a chart of the number of non-conformities per unit from samples of non-constant sizes: a (U) chart.

For the (P) chart, we distinguish between two cases: that in which the sizes of sample batches are identical and that in which they are not. In general, the samples may be of any size and the proportion of defective parts p_i is defined as follows:

$$p_i = \frac{d_i}{n_i} \quad [4.127]$$

These two data are identified for each sample. In the particular case where we have identical samples, i.e. where $\forall i = 1, \ldots, K, n_1 = \ldots = n_K$, the central value, the mean, of the proportions of defective or non-conforming parts is:

$$\overline{p} = \frac{\sum_{k=1}^{K} d_i}{\sum_{k=1}^{K} n} = \frac{\sum_{k=1}^{K} d_i}{n \times K} \quad [4.128]$$

In this case, the limits of the (P) control chart are:

– the upper control limit: $UCL(P) = \overline{p} + 3\sqrt{\frac{\overline{p}(1-\overline{p})}{n}}$;

– the lower control limit: $LCL(P) = \overline{p} - 3\sqrt{\frac{\overline{p}(1-\overline{p})}{n}}$;

– the upper monitoring limit: $UML(P) = \overline{p} + 2\sqrt{\frac{\overline{p}(1-\overline{p})}{n}}$;

– the lower monitoring limit: $LML(P) = \overline{p} - 2\sqrt{\frac{\overline{p}(1-\overline{p})}{n}}$.

In the most general case, where we have different samples, that is:

$$\exists i = 1, \ldots, K, \exists j = 1, \ldots, K, \exists, i \neq j, n_i \neq 0, n_j \neq 0, \Rightarrow n_i \neq n_j \quad [4.129]$$

the central value, the mean, of the proportions of defective or non-conforming parts is:

$$\overline{p} = \frac{\sum_{i=1}^{K} d_i}{\sum_{i=1}^{K} n_i} \qquad [4.130]$$

In this case, the limits of the (P) control chart are the same as those presented previously.

Practically, very specific values are chosen depending on the different sizes of the samples in order to establish a standard chart valid for the different cases such that:

$$0.7\overline{n} \leq n_i \leq 1.3\overline{n} \quad \text{and} \quad \overline{n} = \frac{\sum_{1}^{K} n_i}{K} \qquad [4.131]$$

Then, the limits, with this single size, for the case of a (P) chart, are:

– the upper control limit: $UCL(P) = \overline{p} + 3\sqrt{\frac{p(1-\overline{p})}{\overline{n}}}$;

– the lower control limit: $LCL(P) = \overline{p} - 3\sqrt{\frac{p(1-\overline{p})}{\overline{n}}}$;

– the upper monitoring limit: $UML(P) = \overline{p} + 2\sqrt{\frac{p(1-\overline{p})}{\overline{n}}}$;

– the lower monitoring limit: $LML(P) = \overline{p} - 2\sqrt{\frac{p(1-\overline{p})}{\overline{n}}}$.

The (NP) chart is a graph on which the number of defective parts in each (identically sized) sample is shown as a function of time. It is an attribute-based chart. It still considers identical sample sizes n. The central value, the mean, is then:

$$n\overline{p} = n\frac{\sum_{i=1}^{K} d_i}{\sum_{i=1}^{K} n} = \frac{\sum_{i=1}^{K} d_i}{K} \qquad [4.132]$$

The limits of the (NP) chart are the following:

– the upper control limit: $UCL(NP) = n\overline{p} + 3\sqrt{\frac{p(1-\overline{p})}{\overline{n}}}$;

– the lower control limit: $LCL(NP) = n\overline{p} - 3\sqrt{\frac{p(1-\overline{p})}{\overline{n}}}$;

– the upper monitoring limit: $UML(NP) = n\overline{p} + 2\sqrt{\frac{p(1-\overline{p})}{\overline{n}}}$;

– the lower monitoring limit: $LML(NP) = n\overline{p} - 2\sqrt{\frac{p(1-\overline{p})}{\overline{n}}}$.

The steps in the development of (*P*) and (*NP*) charts are:

Step 1: sampling methods: the size of the sample to be taken, control frequency, number of samples to be taken (20 to 30 samplings for an *NP* chart).

Step 2: calculation of the proportions P and the count of the number of non-conforming units *NP*: calculation of the proportion of non-conformities for each sample. The recording of *np*, the number of non-conforming units in each sample.

Step 3: calculation of the control limits: calculation of the mean proportion of non-conformities. Calculation of *NP*, the mean number of non-conforming units. Calculation of the upper and lower monitoring limits.

Step 4: identification of the control limits on the p and *np* charts: solid horizontal lines for the mean proportion of non-conformities. Dotted horizontal lines for the upper and lower control limits for each chart. The procedure is the same for chart *np*.

Step 5: graphical representation of the P and *NP* charts: the reporting of values, indicating on the ordinates the proportion of non-conformities and on the abscissas the identification of the sample. Connect the points. The same is done for the values of *np*.

Step 6: diagnostics: analysis of the points on each control chart to determine the points exceeding the controls and specific behaviors.

4.6.2.4.11. (C) charts

(C) charts are part of the same family of attribute-based charts. They consist of the graphical represntation of the number of defects per controlled unit (product) as a function of time.

The estimation of the mean of the studied characteristic from the values c_i, representing the number of observed defects for controlled unit i, is:

$$\overline{c} = \frac{\sum_{i=1}^{K} c_i}{K} \quad [4.133]$$

The parameters of the (C) chart are:

– the upper control limit: $\text{CL}_U(C) = \overline{c} + 3\sqrt{\overline{c}}$;

– the lower control limit: $\text{CL}_L(C) = \overline{c} - 3\sqrt{\overline{c}}$;

– the upper monitoring limit: $\text{ML}_U(C) = \overline{c} + 2\sqrt{\overline{c}}$;

– the lower monitoring limit: $\text{ML}_L(C) = \overline{c} - 2\sqrt{\overline{c}}$.

The steps in the development of a (C) chart are:

Step 1: sampling methods: in contrast to p and np charts, the size of the non-conformity control sample constitutes a subgroup of a unit for setting the control frequency with respect to the number of samples to be taken.

Step 2: the counting of non-conformities: the recording of c, the number of non-conformities per controlled unit.

Step 3: calculation of the control limits: calculation of the mean number of non-conformities for k-controlled units; calculation of the upper and lower control limits.

Step 4: identification of the control limits on the p and np charts: solid horizontal lines for the mean number of non-conformities and dotted horizontal lines for the upper and lower control limits.

Step 5: graphical representation of the (C) chart: the reporting of the values with the number of non-conformities as the ordinates and the times of the samples corresponding to each controlled unit as the abscissas. Connect the points.

Step 6: diagnostics: analysis of points exceeding controls.

4.6.2.4.12. Example of an application of control charts

Input data

A company specializes in the manufacture of circular metallic parts. Before implementing a new production system, the company's quality managers wish to perform manufacturing controls concerning the diameters of the manufactured parts.

The company works 16 hours out of 24 with two teams (2×8 h). The controls and sampling are done three times a day. The production rate of the company is 50 parts per hour. Policy dictates that the managers sample a given percentage of the daily production. This percentage equals 3%.

The specifications of the engineering department (ED) state that the diameter of the parts must be 40 mm, allowing a deviation of ± 2.25 mm. The system is considered as efficient.

Calculation of the number of samples to be taken per day and the size of each sample

The company manufactures 50 parts per hour. This means that it has the capacity to produce 800 parts per day (50 parts $\times$ 16 hours). The total number of parts to control (N) is therefore equal to $800 \times 3\%$, i.e. 24 parts to control per day.

Given that there are three samplings per day, the size of each sample (n) equals 24/3, which is eight parts per each sample.

Calculation of the control and monitoring limits of the mean chart

On the basis of input data for this example, the calculation of the control limits of the mean chart can be done according to the estimated mean and standard deviation.

The mean ($\overline{X}$) is given in the assumptions of the problem and is equal to the diameter of the parts as stated by the ED, which is 40 mm. The estimated standard deviation is calculated on the basis of the capability and the tolerance interval (TI) stated by the ED. This interval (TI) is given by the upper tolerance limit ($T_u = 42.25$ mm) and the lower tolerance limit ($T_l = 37.75$ mm), i.e. $TI = 4.5$ mm. The estimated standard deviation (σ^*) equals the TI divided by six times the system's capability ($\sigma^* = TI/(6 \times C_p)$). The system is considered as efficient, and its capability is therefore 1.5. The estimated standard deviation therefore equals $0.5(4.5/(6 \times 1.5))$.

The (upper CL_U and lower CL_L) control limits of the mean chart may now be calculated as shown in equations [4.134] and [4.135]:

$$\text{CL}_U = \overline{X} + (3 \times \sigma^*/\sqrt{n}) = 40.53 \qquad [4.134]$$

$$\text{CL}_L = \overline{X} - (3 \times \sigma^*/\sqrt{n}) = 39.46 \qquad [4.135]$$

The (upper ML_U and lower ML_L) monitoring limits of the mean chart are calculated based on equations [4.136] and [4.137]:

$$\text{ML}_U = \overline{X} + (2 \times \sigma^*/\sqrt{n}) = 40.53 \qquad [4.136]$$

$$\text{ML}_L = \overline{X} - (2 \times \sigma^*/\sqrt{n}) = 39.46 \qquad [4.137]$$

The validity of the control charts

We assume that the mean of the standard deviation observed over the samples taken ($\overline{\sigma}$) is 0.48 and the mean of the ranges ($\overline{R}$) is 1.5. It is now necessary to ensure the validity of the established control charts. To do this, the estimated standard deviation (σ^*) is calculated in two different ways. The first way is based on the mean of the standard deviations (equation [4.138]) and the second way is based on the mean of the ranges (equation [4.139]). The two parameters b_n and d_n in the two equations are read from Table 4.19 according to the size of the sample. Given that the sample size (n) equals 8, this means that b_n equals 0.903 and d_n equals 2.847:

$$\sigma^* = \overline{\sigma}/b_n = 0.531 \qquad [4.138]$$

$$\sigma^* = \overline{R}/d_n = 0.526 \qquad [4.139]$$

The difference between the two values is minimal (0.88%), which allows us to confirm the validity of the control charts.

The distribution followed by the parts and the means measured over each sample

The probability distribution that the parts follow is a normal distribution with mean $\overline{X} = 40$ and standard deviation $\sigma^* = 0.5$: $N(\overline{X}, \sigma^*)$. The distribution followed by the means measured over each sample is also a normal distribution, with mean $\overline{X} = 40$ and standard deviation $\sigma^* / \sqrt{n} = 0.1768$: $N(\overline{X}, \sigma^*/\sqrt{n})$.

Manufacturing control of eight samples

Taking into account the measurements of eight parts ($X1 \rightarrow X8$) from the four samples ($S1 \rightarrow S4$) shown in Table 4.22, we will now perform a manufacturing control of these four samples with a mean chart.

	$X1$	$X2$	$X3$	$X4$	$X5$	$X6$	$X7$	$X8$
$S1$	41	40	40	39	42	40	38	40
$S2$	41	42	42	40	43	42	42	41
$S3$	37	40	39	43	41	41	40	40
$S4$	43	40	42	38	39	40	39	41

Table 4.22. *The observations over the four samples to be controlled*

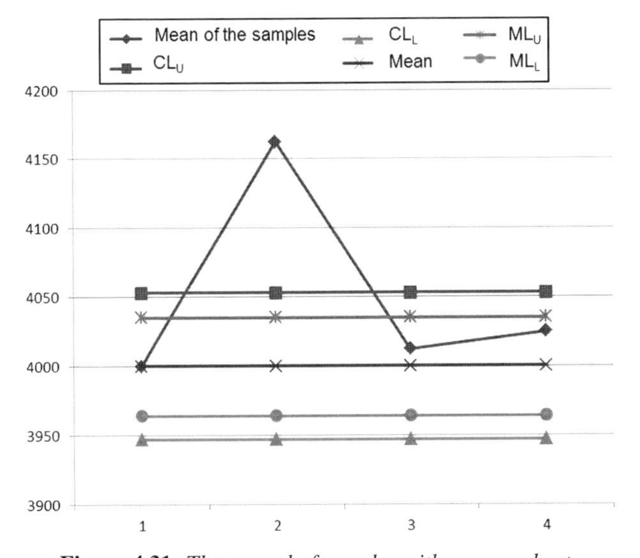

Figure 4.31. *The control of samples with a mean chart*

The first step consists of calculating the mean of the measurements of each sample: $\overline{S1} = 40$, $\overline{S2} = 41.625$, $\overline{S3} = 40.125$ and $\overline{S4} = 40.25$. From these means, we may now use the control (CL_U and CL_L) and monitoring (ML_U and ML_L) limits calculated already to verify the conformity of each sample. The means of the four samples are compared with the different limits. This comparison is done by putting the mean of the four samples on the control chart as shown in Figure 4.31. By carrying out manufacturing control, we find that the mean of sample S2 exceeds the upper control limit (CL_U). The mean of the three other samples is found between the control and monitoring limits, which validates the conformity of samples S1, S3 and S4. The non-conformity of sample S2 is therefore probably due to an incident in the manufacturing process.

Chapter 5

Scheduling

5.1. Introduction

The main activity of an industrial company in the short term is to organize its production on this time frame. Scheduling production orders involves deciding which machine(s) to be used and in what order to achieve the production. Because of the diversity in production processes and management styles, which vary between industries, numerous methods allowing the optimization of task allocation have been developed.

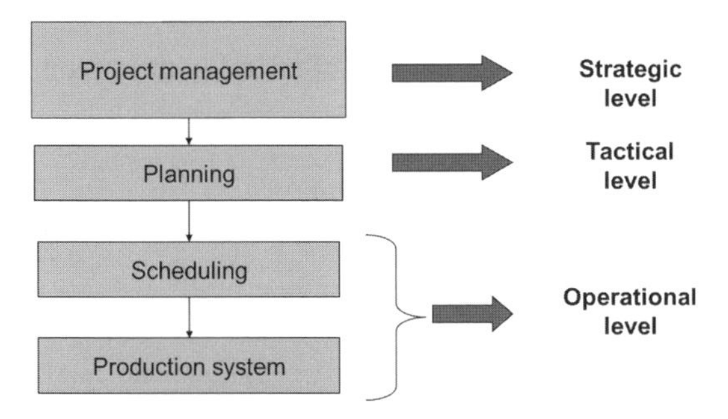

Figure 5.1. *The different time frames*

In the discussion of the basic notions of scheduling problems and the nature of these problems, we present the main notation and evaluation criteria. Then, project scheduling problems are addressed, followed by the discussion on single-machine problems. Finally, flow shop and parallel machine workshop scheduling problems are presented.

5.2. Scheduling problems

5.2.1. *Basic notions*

In a scheduling problem, four fundamental concepts are involved: tasks (jobs), resources, constraints and objectives.

A job is defined by a set of operations that are to be executed. It is characterized in time by a start and end date.

A resource is a machine or a human involved in the performance of work. Constraints represent limits in time, technology or resources. Objectives are the criteria to be optimized with regard to time, resources, costs or output.

Scheduling problems may be divided into two categories depending on the number of operations necessary to carry out jobs: problems for which each job requires a single operation (they vary according to the configuration of the considered machines – single machines, dedicated machines or parallel machines) and problems for which each job requires several operations. The latter are called "workshop problems" and vary according to the way in which operations are ordered on different machines (open shop, flow shop or job shop).

5.2.2. *Notation*

We mainly use the following notations in the majority of problems:

– C_i represents the end date (the completion date) of job i;

– r_i represents the release date (the date of arrival to the system) of job i;

– d_i represents the due date (the required date) of job i;

– p_i represents the processing time of job i.

5.2.3. *Definition of the criteria and objective functions*

There are several criteria that allow us to evaluate the quality of solutions as a function of a company's objectives and priorities. These criteria define different objective functions and therefore lead to the elaboration of different mathematical problems to be resolved. A criterion rates a solution before qualifying it. Let us take the following example of four jobs with different processing times to be carried out on one machine.

Different criteria may be used to evaluate the quality of the solutions. The most common are the time presence in the system (flow time), the deviation from the due date (lateness), the true delay (tardiness) or even the earliness.

5.2.3.1. *Flow time*

The flow time measures the time that a job (or the manufacturing order) takes in the workshop. It is denoted by F_i:

$$F_i = C_i - r_i \quad [5.1]$$

Figure 5.2 shows that the flow time calculation is $4 + 7 + 9 + 10 = 30$.

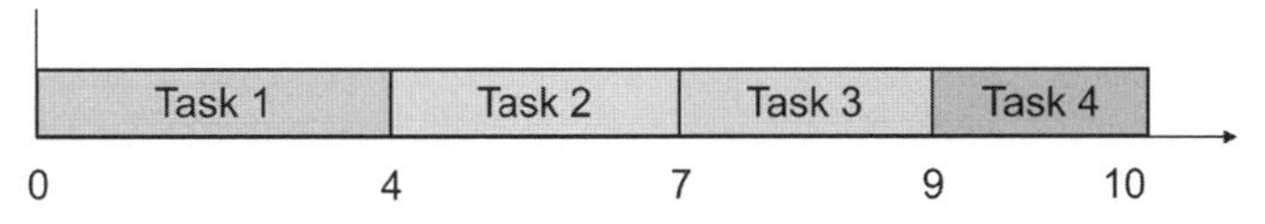

Figure 5.2. *An example of calculating F_i*

Flow time minimization reduces the stock of works in progress in the workshop.

5.2.3.2. *Lateness*

The lateness may be favorable (negative) or unfavorable (positive). It is denoted by L_i and calculated in the following way:

$$L_i = C_i - d_i \quad [5.2]$$

In the example in Figure 5.3, the lateness is $((4-5)+(7-3)+(9-10)+(10-4)) = (-1) + 4 + (-1) + 6 = 8$. Its minimization leads to the completion of jobs as early as possible.

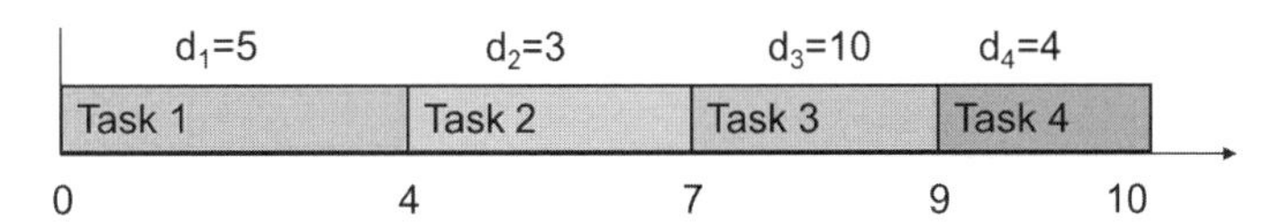

Figure 5.3. *An example of calculating L_i*

5.2.3.3. *Tardiness*

The tardiness measures the delay in a job. Delay is calculated if the job is late, and is zero otherwise; hence its expression is as follows:

$$T_i = \max\{0, L_i\} \quad [5.3]$$

Taking again the example in Figure 5.3, the tardiness is $(0 + (7 - 3) + 0 + (10 - 4)) = 4 + 6 = 10$.

5.2.3.4. *The earliness*

The earliness is calculated if a job finishes before the delay d_i, and is zero otherwise:

$$E_i = \max\{0, -L_i\} \qquad [5.4]$$

In the context of the example in Figure 5.3, the earliness is $(-(4 - 5) + 0 + (-9) + (-10)) + 0 = 1 + 1 = 2$. The minimization of this criterion avoids finishing of jobs before their due dates.

5.2.3.5. *Objective functions*

Objective functions may be defined based on certain criteria. The best studied are the following.

5.2.3.5.1. Minimization of the total flow time

The minimization of the sum of the jobs' completion times allows the reduction in works in progress, hence the criterion is written as follows:

$$\sum_{i=1}^{n} F_i \qquad [5.5]$$

where F_i is the flow time of job i. The minimization of this criterion leads to the reduction in the mean of the flow times:

$$\frac{\sum_{i=1}^{n} F_i}{n} \qquad [5.6]$$

5.2.3.5.2. Minimization of the total tardiness

The minimization of the sum of the tardinesses is an important criterion. Tardiness may lead to customer dissatisfaction and financial losses:

$$\sum_{i=1}^{n} T_i \qquad [5.7]$$

where T_i is the tardiness of job i. The minimization of this criterion is equivalent to the minimization of the tardiness average:

$$\frac{\sum_{i=1}^{n} T_i}{n}$$

5.2.3.5.3. Minimization of the total earliness

The minimization of the sum of the earlinesses is an objective function allowing the reduction in stocks, enabling just-in-time production:

$$\sum_{i=1}^{n} E_i \qquad [5.8]$$

where E_i represents the earliness of job i.

5.2.3.5.4. Minimization of the makespan

The minimization of the end date of the last job is also known as *makespan* minimization. The makespan represents the time separating the input of the first product into the system and the output of the final product.

This criterion measures productivity. In effect, performance increases as this time decreases:

$$C_{\max} = \max_{1 \leq i \leq n} C_i \qquad [5.9]$$

where C_i represents the end date (the completion time) of job i.

5.2.3.5.5. Minimization of the number of tardy jobs

The minimization of the number of delayed jobs is widely used in agribusiness. Expiration dates require that delays do not accumulate, so that the product may be sold within the statutory period:

$$N_t = \sum_{i=1}^{n} U_i \qquad [5.10]$$

where:

$$U_i = \begin{cases} 0 & \text{if } C_i \leq d_i \\ 1 & \text{otherwise} \end{cases} \qquad [5.11]$$

where C_i is the completion time of job i and d_i is its due date.

5.2.3.5.6. Graham *et al.* notation

In 1979, Graham *et al.* [GRA 79a] proposed a notation with three fields to identify the different scheduling problems:

$$\alpha/\beta/\gamma \qquad [5.12]$$

The α field provides information about the resource, β gives indications about the constraints and finally γ tells us about the criteria to be minimized. This notation is often used to designate considered scheduling problems.

5.2.3.6. *Properties of schedules*

Scheduling problems are characterized by, among other things, the type of their criteria. A criterion may be regular or irregular.

5.2.3.6.1. Regular criteria

A criterion is regular if its value decreases with the decrease in at least one of the jobs' completion times.

Regular criteria are decreasing functions of the completion times of the operations.

5.2.3.6.2. Irregular criteria

An irregular criterion is a non-regular criterion.

5.2.3.6.3. Semi-active scheduling

A schedule is semi-active if no operation may begin earlier without modifying the order in which the operations are carried out on the machines.

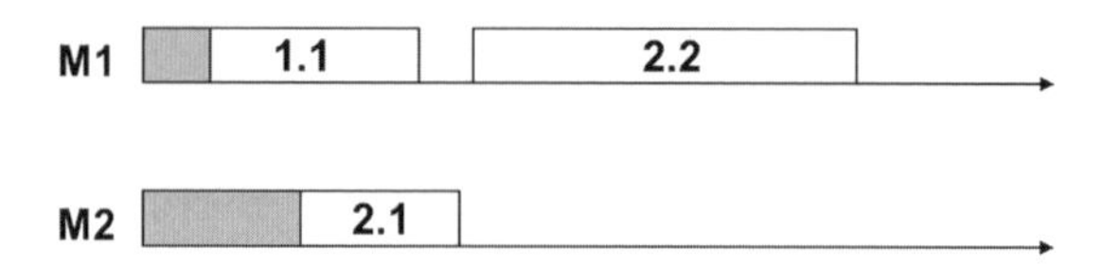

Figure 5.4. *An example of semi-active scheduling*

In the example in Figure 5.4, we can see that operation 2.2 cannot begin earlier as it must wait for operation 2.1 to finish. If operation 2.2 began earlier, then the order of operations on the machines would be modified.

5.2.3.6.4. Non-semi-active scheduling

A schedule is non-semi-active if it is possible to begin one or more operations earlier without changing the order in which the operations are performed on the machines.

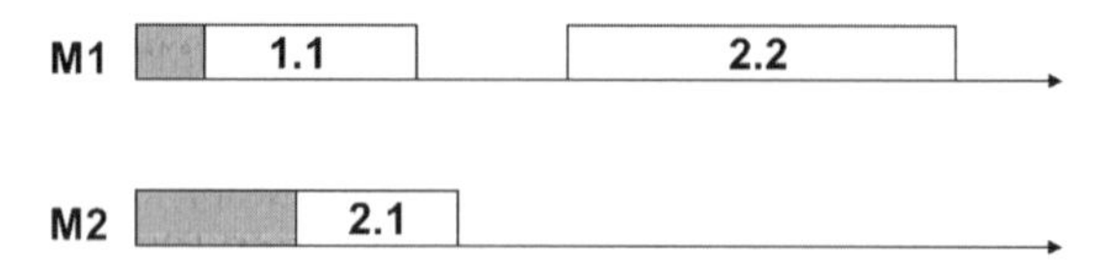

Figure 5.5. *An example of non-semi-active scheduling*

In the example in Figure 5.5, we can see that operation 2.2 may begin earlier as it must wait for operation 2.1 to have finished, which is the case. If operation 2.2 began earlier, then the order of operations would be interchanged.

5.2.3.6.5. Active scheduling

A schedule is active if no operation may begin earlier without delaying at least one other. An active schedule is also semi-active.

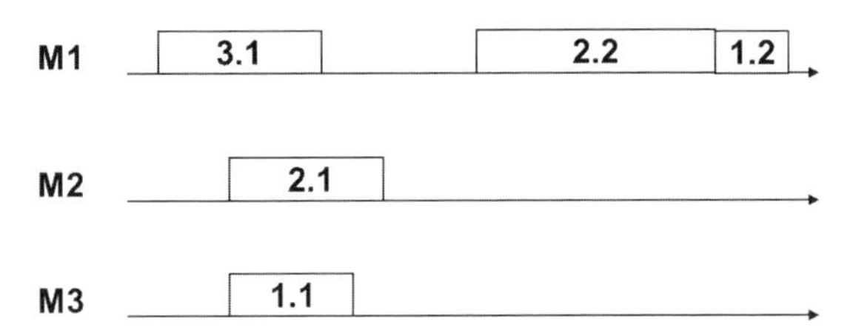

Figure 5.6. *An example of non-active scheduling*

In the example in Figure 5.6, we can see that operation 1.2 may begin earlier as it must wait for operation 1.1 to have finished, which is the case. If operation 1.2 began earlier, then the order of operations would be interchanged. This schedule is therefore not active.

5.2.3.6.6. Dominant scheduling

The sets of semi-active and active schedules are dominant for regular criteria. This means that to optimize a regular criterion, it is sufficient to focus on active and semi-active schedules.

5.2.4. *Project scheduling*

5.2.4.1. *Definition of a project*

A project is a set of operations (or jobs) that allows the achievement of an objective that is clearly expressible and presents a certain quality of uniqueness. For example, the construction of a factory is a project.

A job is an operation that is carried out to achieve the initially set objective. For example, the construction of the foundation of a house is a job. Performing a job requires a certain number of resources.

A resource is a financial, material or human means necessary for the completion of a job. Resources may be of two types:

– consumable: those consumed during the operation of a job;

– renewable: those consumed but that become available again.

For example, consumable resources may be money, raw materials, etc. Renewable resources may be machines, operators, etc.

Jobs are subjected to constraints. There are several types of constraints for the management of a project:

– Precedence constraints: job A may not begin before job B has finished.

– Constraints of time localization: a job may only be performed during a specific time interval.

– Disjunctive constraints: two jobs A and B cannot be simultaneously carried out because they use the same renewable resource.

– Cumulative constraints: the jobs require the presence of resources to be consumed, and these resources are in limited quantities, but greater than one.

5.2.4.2. *Projects with unlimited resources*

When there are no particular constraints on the resources, then the project is said to be of unlimited resources. There are several graphical representations used to solve the problem of scheduling the jobs involved in the project.

5.2.4.2.1. Potential event representations

The graph for the Program Evaluation and Review Technique (PERT) shows arcs (jobs) and vertices (events). This graph is oriented in the direction of the passage of time. In effect, it goes from one "event" point, representing the start or end of a job, to an "event" point of the same type. Let us take the example of a project composed of nine jobs whose precedence relations and durations in days are given in Table 5.1.

Job	Preceding jobs	Duration
A	-	4
B	A	6
C	-	4
D	-	11
E	B, C	10
F	B, C	24
G	A	6
H	D, E, G	10
I	F, H	10

Table 5.1. *An example of a project composed of nine jobs*

The representation of this example is given in Figure 5.7.

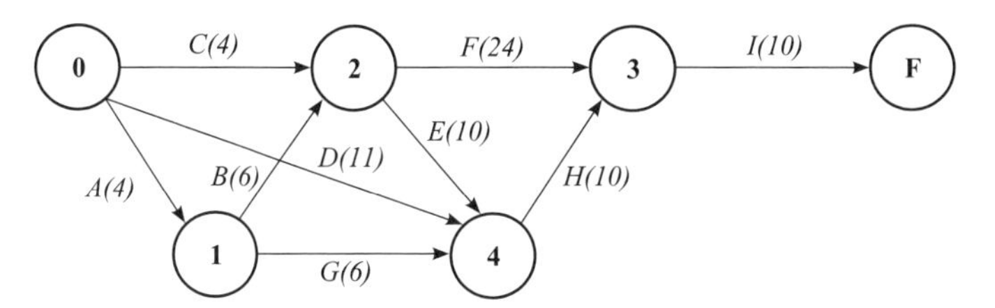

Figure 5.7. *A PERT representation*

5.2.4.2.2. Potential job representation

This representation was proposed after the introduction of the PERT. It is composed of vertices representing jobs and arcs representing constraints. The value shown on each arc corresponds to the minimum time which must elapse between two jobs. Furthermore, an arc between two points represents a precedence constraint between two jobs.

The potential job representation of the example is given in Figure 5.8.

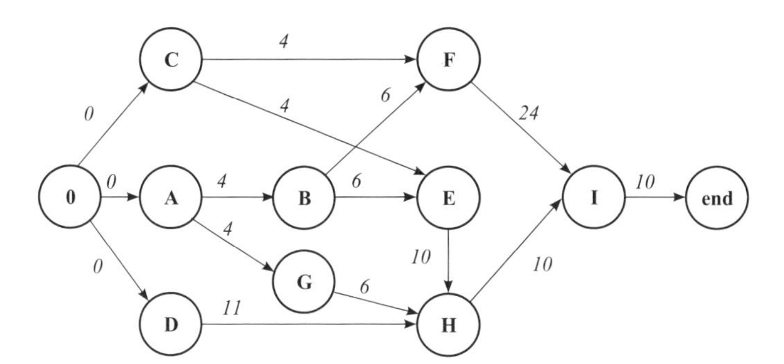

Figure 5.8. *A potential job representation*

5.2.4.2.3. Resolution methods

The aim is to propose a job schedule for the project that finishes as early as possible. In this case, the schedule is said to be left-justified. This arrangement has a single rule, which is that all jobs be scheduled as early as possible.

From this schedule, we may also propose a right-justified schedule, i.e. a tardy schedule. This schedule may be the latest optimal alignment or the latest alignment with a given delay. When the schedule is the latest optimal schedule, the last job finishes as early as possible, i.e. at the same time as for the earliest possible schedule. Whereas a tardy schedule with a given delay requires the last job to finish at a specified time, which is greater than or equal to the end time of the earliest possible schedule, taking account of the delay.

5.2.4.2.4. The Dijkstra method

The Dijkstra method [PRI 97] allows the calculation of an earliest possible schedule. This method dates back to 1959. It is of polynomial-time complexity, and allows us the determination of the shortest path through a graph representing a network between towns or a sequence of steps for a project.

For this algorithm to be applicable, the graph must have positive arc lengths and no circuits, which avoids problems of negative absorbing circuits.

We write:

– λ_i: the label of job i, providing information about its earliest start date, i.e. the shortest path length between node 0 ($\lambda_0 = 0$) and node i. The labels of the nodes $i \neq 0$ are initialized as ∞. Once the value of a label is definitively updated, it is said to be fixed;

– P_i: the set of predecessors of i whose label is fixed;

– a_{ij}: the value of the arc between nodes i and j on the potential graph.

The steps of the algorithm are then the following:

– initialize the labels $\lambda_i = \infty\ \forall i \neq 0$, and $\lambda_0 = 0$;

– seek the node x whose value is the minimum among the nodes whose label is not fixed. If such a node does not exist, the algorithm terminates;

– fix the label of node x;

– adjust the label of each successor and start again from step 2.

Let us continue with the example in Table 5.2. The fixed labels are shown on the graph at each step (Figure 5.9).

The nodes 0, A, C and D have a fixed label, then among the nodes whose predecessors are all fixed, we choose to fix the label of node B (Figure 5.9). Each time, we update the labels of the successors of the fixed nodes. This then leads us to fix node G then E and F. Next, H and I are fixed (Figure 5.10).

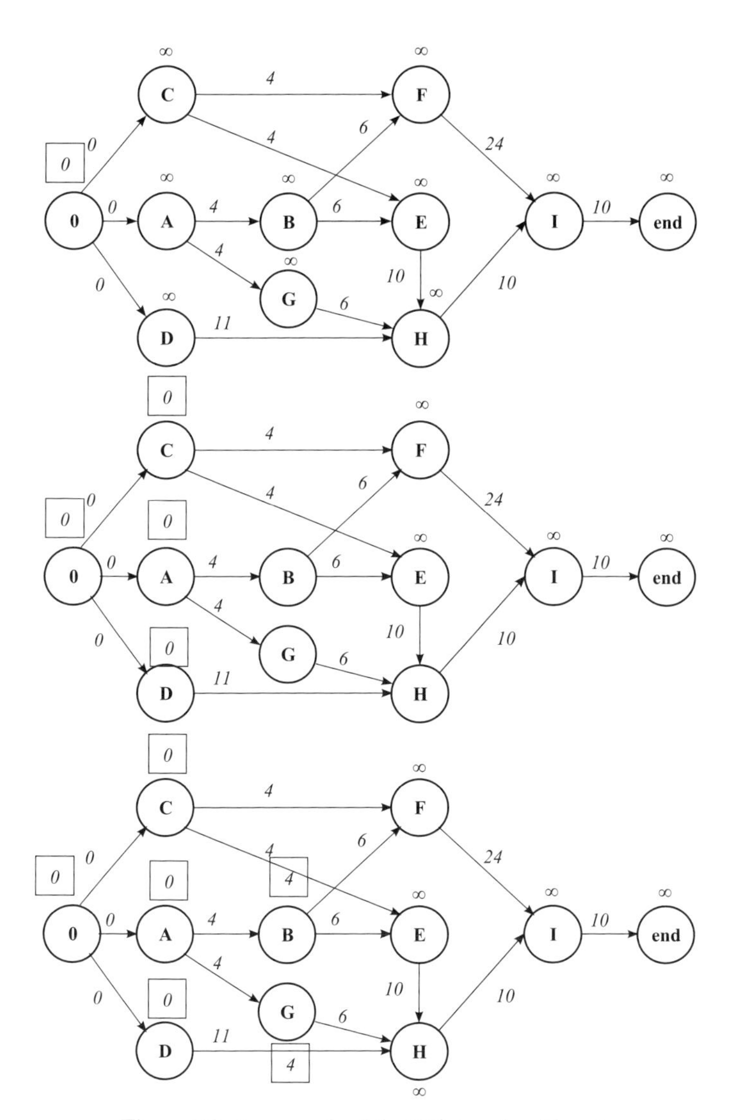

Figure 5.9. *An example of the Dijkstra algorithm -1*

We then obtain the following schedule, with an earliest end at a time of 44 units (Figure 5.11).

The Dijkstra algorithm may be adapted and applied to obtain a tardy schedule [PRI 97]. In this case, the last job is scheduled at the earliest and resources are used as late as possible.

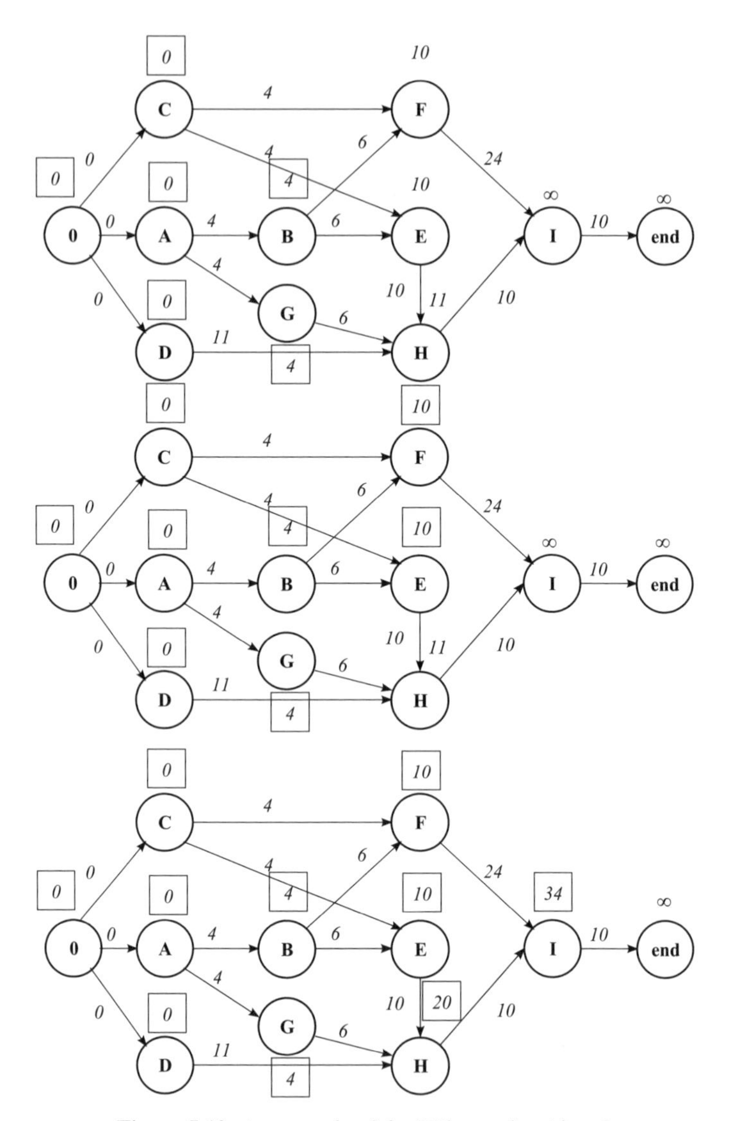

Figure 5.10. *An example of the Dijkstra algorithm -2*

The notation employed is as follows:

– θ_i: the label of job i, giving information about its latest start date;

– S_i: the set of fixed successors of i;

– a_{ij}: the value of the arc between nodes i and j of the potential graph.

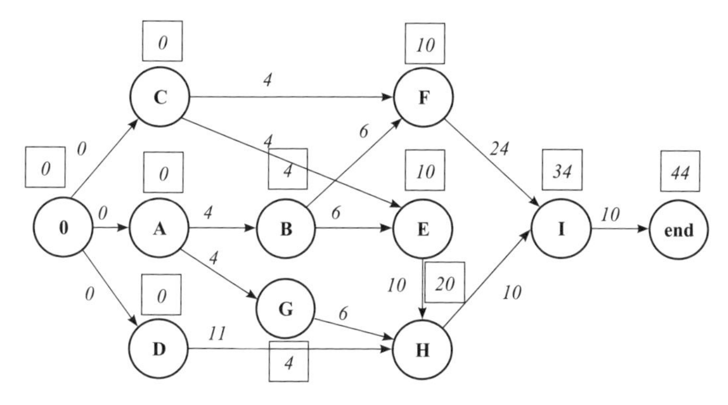

Figure 5.11. *An example of the Dijkstra algorithm -3*

The steps of the algorithm are then the following:

– Initialize the labels $\theta_i = 0$, except for $\theta_{end} = \Delta$, where Δ corresponds to the earliest end date, which was found previously.

– Seek the node x whose value is maximal among the nodes whose label is not fixed. If such a node does not exist, then the algorithm terminates.

– Fix the label of node x.

– Adjust the label of each predecessor and begin again from step 2.

Applying this algorithm to the previous example, we obtain the first graph (Figure 5.12), and then fix nodes I and H. After this, we fix nodes G and E, and then D, updating at each iteration the labels of the predecessors (Figure 5.13). Finally, we fix the labels of nodes F, A and C (Figure 5.14).

The application of this algorithm highlights the critical path, that is the set of jobs for which there is no margin (i.e. those for which the earliest date is the same as the latest) and the sum of whose durations sets the duration of the project. The example is shown in Figure 5.15.

5.2.4.2.5. The Bellman algorithm

The Bellman algorithm [PRI 97] allows the shortest path to be found through a circuit-less graph which takes negative as well as positive costs. This method is based on dynamical programming and on a decision-level decomposition approach.

t_i denotes the label of node i. We intialize $t_i = 0$ for all i from 1 to n. b_{ij} is the cost of the arc going from i to j. We then obtain:

$$t_j = \min_{i \in P_j} t_i + b_{ij} \qquad [5.13]$$

where P_j is the set of predecessors to node j.

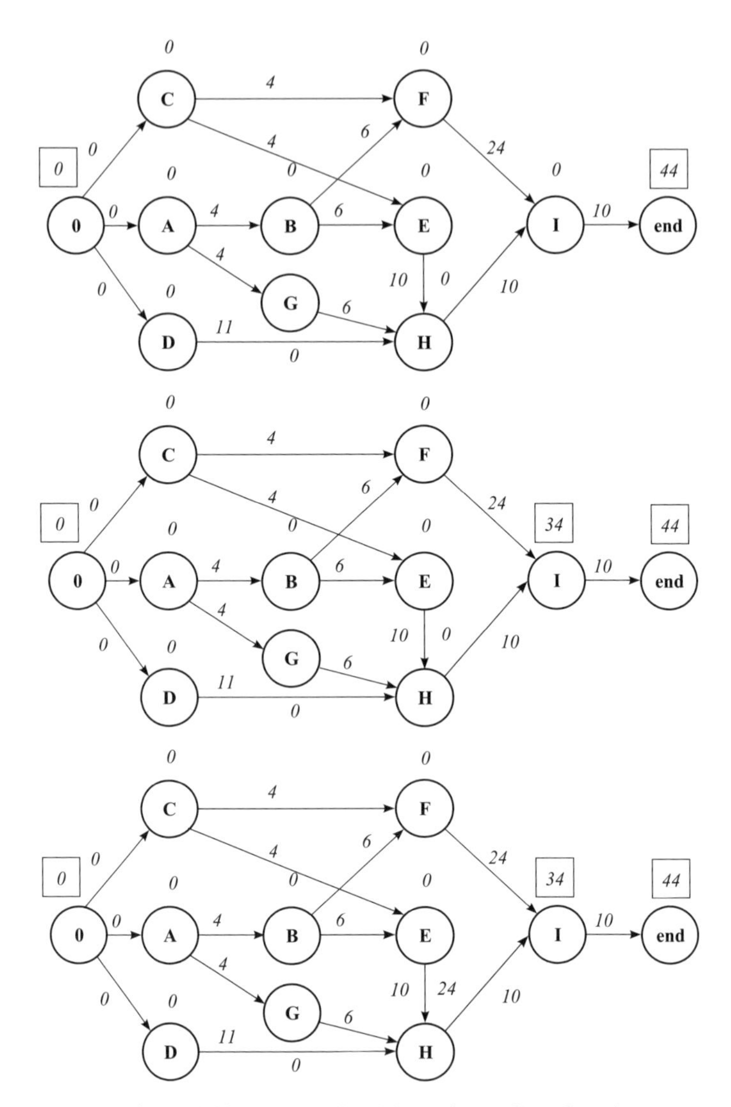

Figure 5.12. *An example of the Dijkstra algorithm -4*

Let us consider the graph in Figure 5.16. It is, first of all, necessary to reformat the graph and decompose it into levels. We fix the label of node 3 at 0 (Figure 5.17) and then move onto the next level (Figure 5.18).

We obtain the graph in Figure 5.19. Next we arrive at the last level as shown in Figure 5.20.

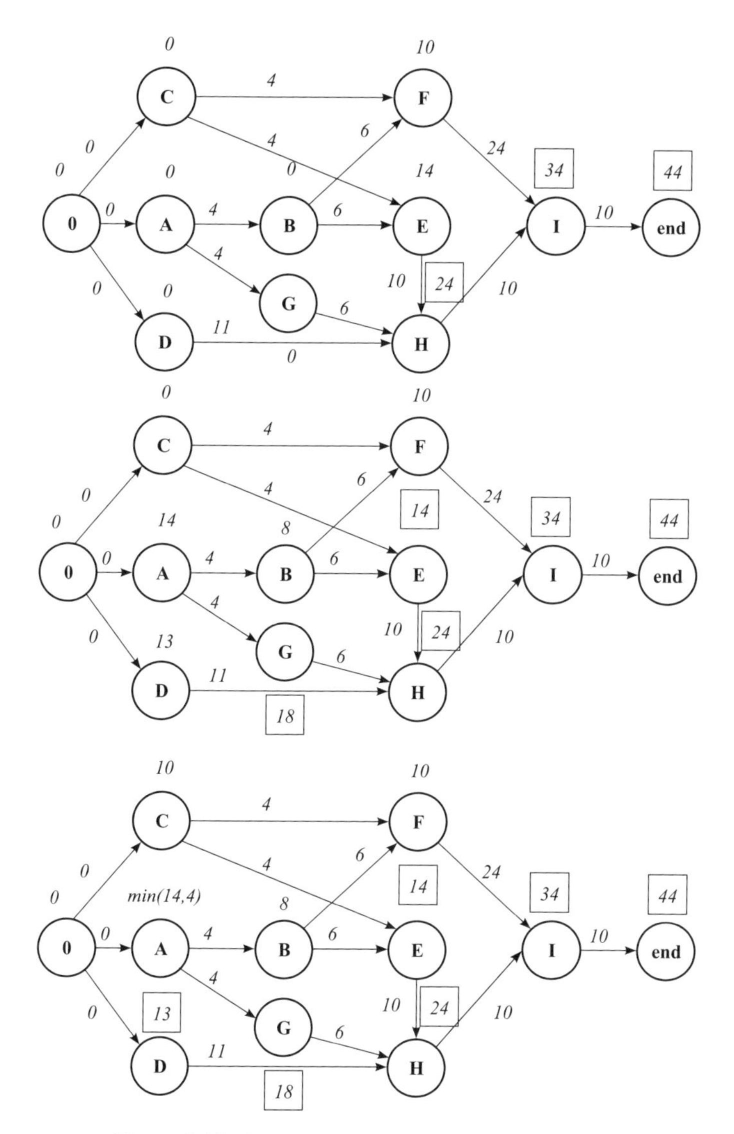

Figure 5.13. *An example of the Dijkstra algorithm -5*

5.2.4.3. *Projects with consumable resources*

When we take into account the consumption of resources in order to establish a feasible earliest possible scheduling, we use the PERT. We use information about the resource consumption per job and the times at which the resources become available.

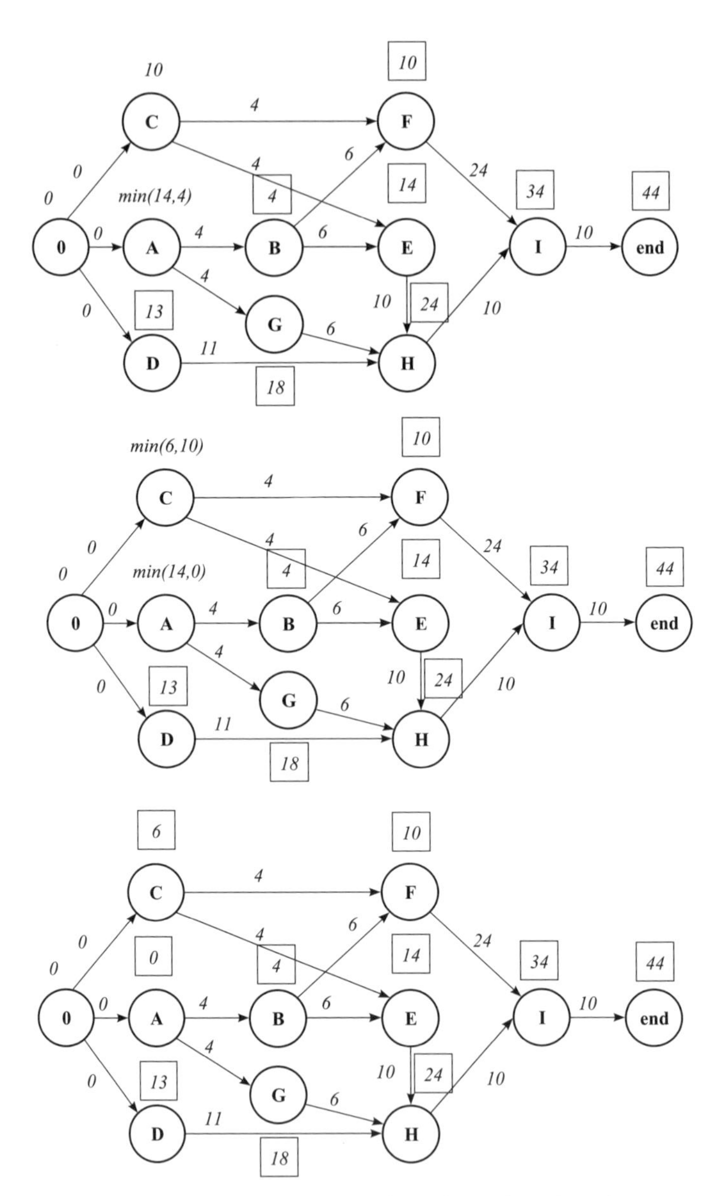

Figure 5.14. *An example of the Dijkstra algorithm -6*

S_i denotes the quantity of the considered consumable resources necessary for the completion of job i and τ_i denotes the difference between the instant when the resources are available and the start of job i.

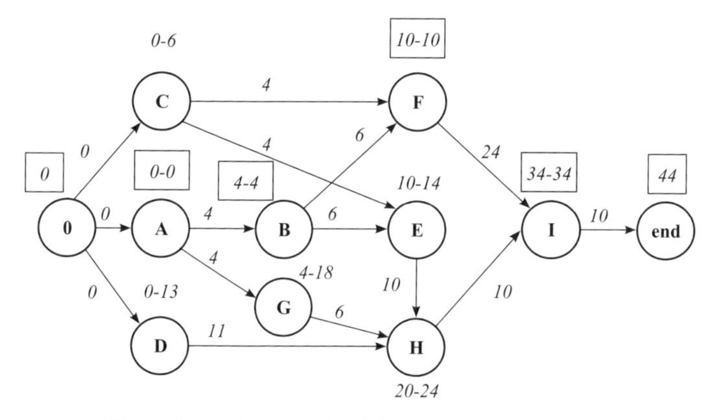

Figure 5.15. *An example of the Dijkstra algorithm -7*

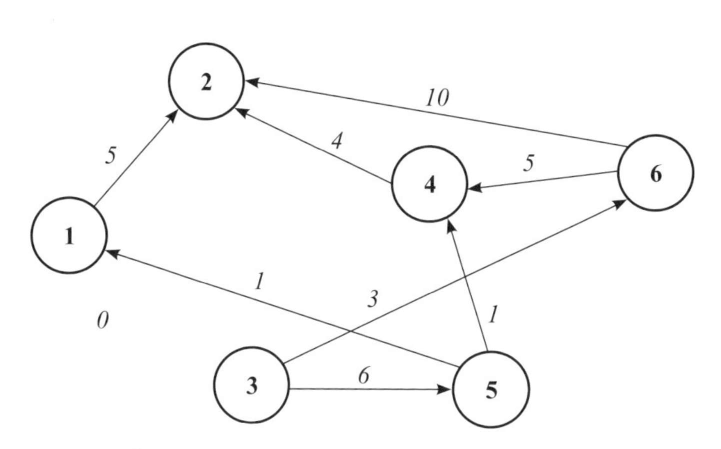

Figure 5.16. *An example of the Bellman algorithm*

e_k denotes the time of delivery k of the resource and S_k denotes the quantity delivered at time k.

5.2.4.3.1. Resolution method

The different steps of the method of resolution are:

– Determine the earliest possible schedule. Calculating the earliest schedule gives us the earliest possible time λ_n at which we may finish a job, without taking account of the limitations due to the consumable resource.

– Determine the latest possible schedule. Calculating the latest schedule lets us finish a job as early as possible while using the consumable resource as late as possible.

– Calculate, as a function of time, the cumulative resource–consumption curve for the latest schedule and the cumulative input curve for the resource.

– Compare the two curves and determine the length δ_i of the intervals where consumption is greater than the available quantity.

– Shift the start dates of the job in the latest schedule by the quantity $\max \delta_i$.

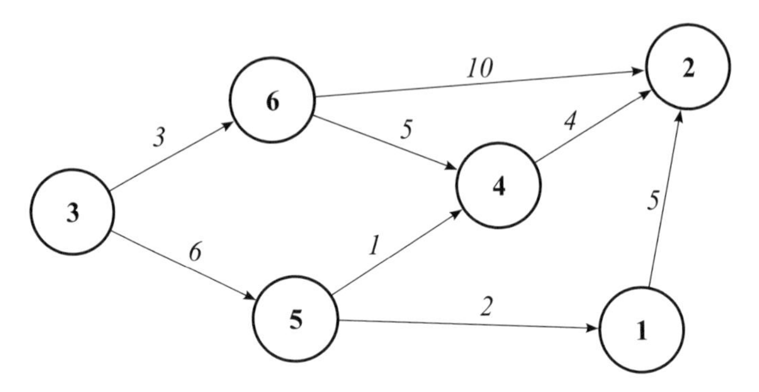

Figure 5.17. *An example of the Bellman algorithm -1*

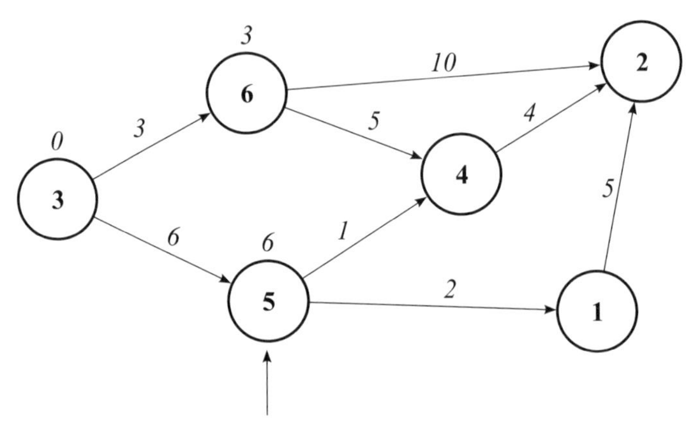

Figure 5.18. *An example of the Bellman algorithm -2*

EXAMPLE 5.1.– Let us consider the example given in Table 5.2, indicating the predecessors, the quantity of resources required and the duration for each job.

Table 5.3 gives the delivery date of each resource, as well as the corresponding delivered quantity.

The graphical model of the project is illustrated in Figure 5.21.

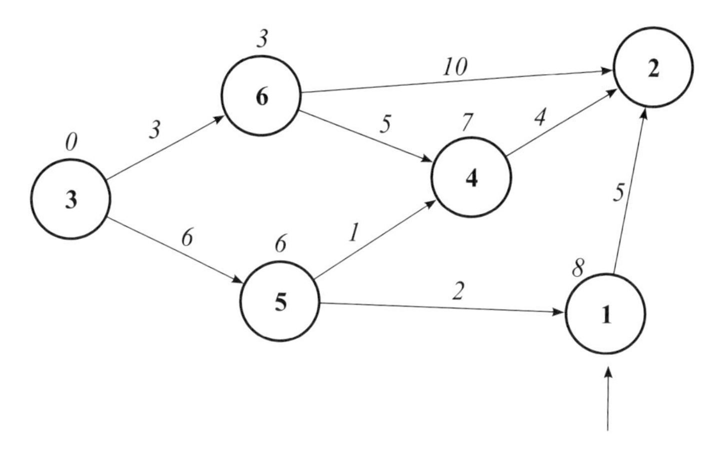

Figure 5.19. *An example of the Bellman algorithm -3*

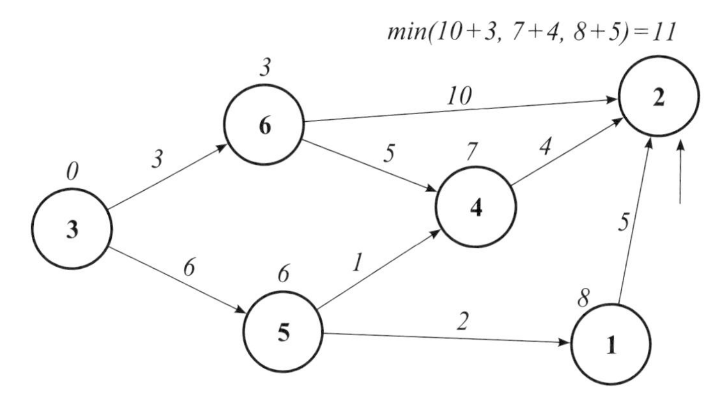

Figure 5.20. *An example of the Bellman algorithm -4*

i	1	2	3	4	5	6	7	8	9	10	11	12	13	14	15
P_i		1	1	1	2	2, 3	2	3, 4	6, 7, 8	7	2	9	9, 10, 11	5	12, 13, 14
s_i	0	10	18	14	8	4	16	8	18	12	10	10	6	16	0
d_i	0	3	6	8	7	2	4	5	1	3	9	4	2	3	0

Table 5.2. *The data for Example 5.1*

Table 5.4 shows the earliest start dates, latest start dates (ensuring the earliest end dates) and the latest end dates.

Table 5.5 shows the cumulative quantities of consumed and delivered resources, and thus highlights the problems occurring due to the lack of resources.

k	1	2	3	4
e_k	5	10	15	20
S_k	40	40	40	40

Table 5.3. *The results from the data for Example 5.1*

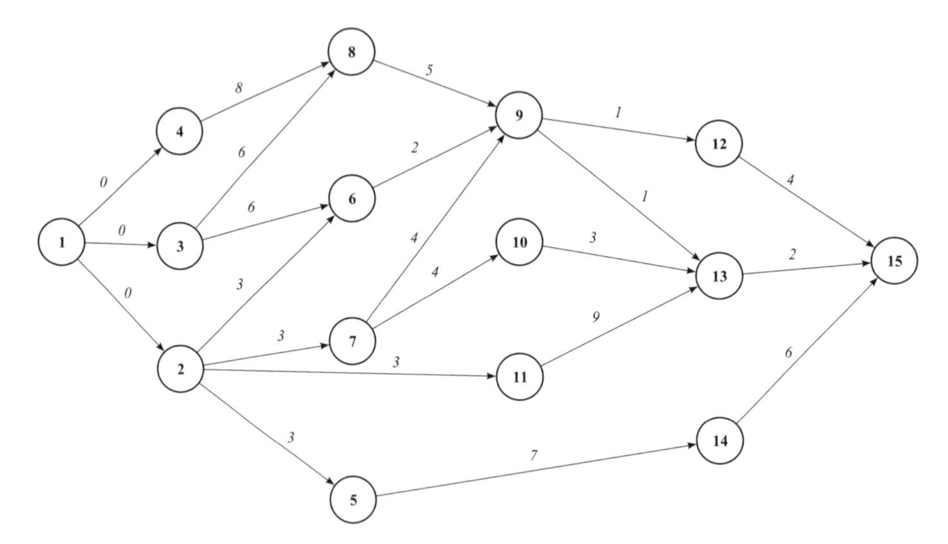

Figure 5.21. *The graphical representation*

i	2	3	4	5	6	7	8	9	10	11	12	13	14	15
λ_i	0	0	0	3	6	3	8	13	7	3	14	14	10	8
θ_i	2	2	0	5	11	9	8	13	13	7	14	16	12	18
$\theta_i + d_i$	5	8	8	12	13	13	13	14	16	16	18	18	18	18

Table 5.4. *The schedule for Example 5.1*

t	5	8	10	12	13	14	15	16	18	20
$\sum S(t)$	40	40	80	80	80	80	120	120	120	160
$\sum s_i$	10	42	42	50	58	96	96	118	150	150

Table 5.5. *The consumption of resources in Example 5.1*

5.2.4.4. *Minimal-cost scheduling*

In the project scheduling problem, there is the possibility to reduce the duration of jobs given a certain cost. The applied method is known as "PERT/cost".

The approach consists of first generating the PERT graph associated with the problem, and then seeking the critical path. It is composed of the jobs that form this path responsible for the duration of the project. We should therefore focus on the jobs if we wish to reduce the duration of the project. The reduction in the duration of the jobs should be done step-by-step, as the critical path is susceptible to change with each modification.

Job	Precedence	Normal duration (days)	Accelerated duration (days)	Daily cost
A		6	4	8
B		8	6	10
C	A	4	1	6

Table 5.6. *The data for Example 5.2*

EXAMPLE 5.2.– Let us consider the following example showing three jobs whose characteristics are reported in Table 5.6. The corresponding PERT graph is shown in Figure 5.22.

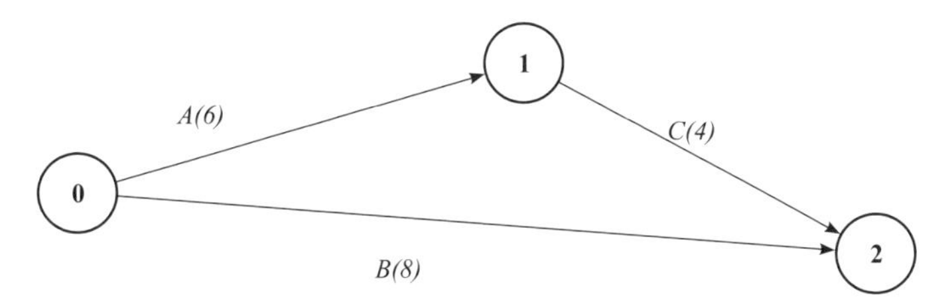

Figure 5.22. *A PERT/cost example*

The critical path is shown by jobs A and C (Figure 5.23).

We therefore have the choice of reducing either job A or job C to reduce the duration of the project. We reduce job C by two units, as it has the smallest cost (cost = 6). We obtain the graph in Figure 5.24.

All of the jobs are now critical. If we wish to continue reducing the duration of the project, it is necessary to reduce both paths. We may only reduce C by a single unit. We will therefore reduce B and C by a single unit. The cost equals 10 + 6 = 16. The total cost equals 12 + 16 = 28. We now have the graph in Figure 5.25.

We may still reduce A and B by one unit. The cost equals $8 + 10 = 18$. The total cost is $28 + 18 = 46$. We obtain the solution in Figure 5.26.

B and C cannot be reduced further. It is useless to reduce A as this does not reduce the duration of the project since B is critical.

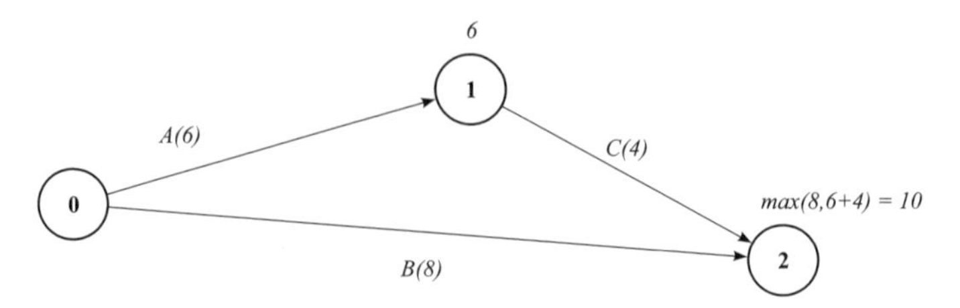

Figure 5.23. *A PERT/cost example -(1)*

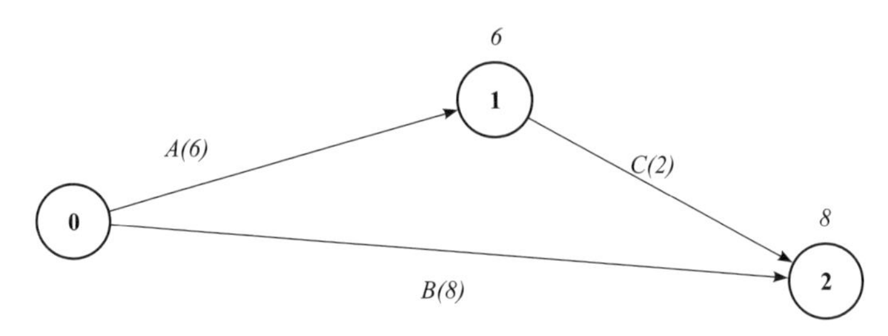

Figure 5.24. *A PERT/cost example -(2)*

5.2.5. *Single-machine problems*

The study of single-machine problems forms the basis of the study of workshop scheduling problems. Even though there are few single-machine production systems, their study enables us to find properties and ideas for the resolution of more complicated problems. These results can also be applied for the focus on bottleneck machines.

As mentioned earlier, there are several criteria, that is to say several ways to evaluate a schedule. Depending on the considered criterion, the approach will be different.

5.2.5.1. *Minimization of the mean flow time* $1/r_i = 0/\sum C_i$

This problem consists of minimizing $\sum F_i$ or $\sum F_i/n$ given that all of the jobs are available (or arriving in the system at the same time). The minimization of this criterion is equivalent to the minimization of $\sum C_i$.

In the case where $r_i = 0 \; \forall i = 1...n$, the problem is polynomial, which is not the case for $1/r_i/\sum C_i$ which itself is NP-hard. The method used is the shortest processing time (SPT) algorithm. This consists of scheduling jobs in increasing order of their processing times p_i. Let us consider two jobs i and j such that $p_i > p_j$ (Figure 5.27).

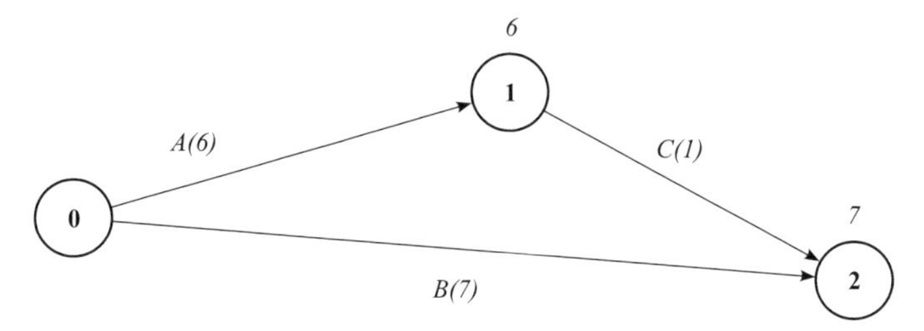

Figure 5.25. *A PERT/cost example -(3)*

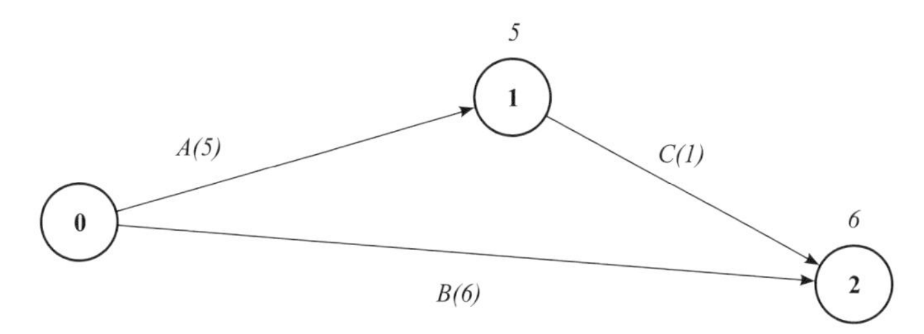

Figure 5.26. *A PERT/cost example -(4)*

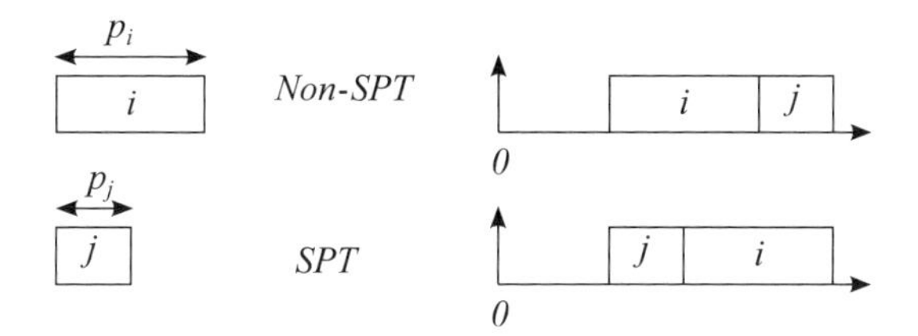

Figure 5.27. *The SPT method*

Following the SPT method, we schedule job j, and then job i. To show this, we proceed by permutation of the jobs. We show that if we consider the scheduling of n jobs, by switching jobs i and j in the order of the SPT, without changing the others, we improve the solution. Based on Figure 5.27, we have:

$$F_{\text{NON-SPT}} = \frac{\sum_{l=1,\neq i,j}^{n} F_l + F_i(\text{non}-\text{SPT}) + F_j(\text{non}-\text{SPT})}{n} \quad [5.14]$$

$$F_{\text{SPT}} = \frac{\sum_{l=1,\neq i,j}^{n} F_l + F_i(\text{SPT}) + F_j(\text{SPT})}{n} \quad [5.15]$$

From here, we calculate:

$$F_{\text{NON-SPT}} - F_{\text{SPT}} = \frac{\sum_{l=1,\neq i,j}^{n} F_l + F_i(\text{non}-\text{SPT}) + F_j(\text{non}-\text{SPT})}{n} - \frac{\sum_{l=1,\neq i,j}^{n} F_l + F_i(\text{SPT}) + F_j(\text{SPT})}{n} \quad [5.16]$$

$$F_{\text{NON-SPT}} - F_{\text{SPT}} = \frac{F_i(\text{non}-\text{SPT}) + F_j(\text{non}-\text{SPT})}{n} - \frac{F_i(\text{SPT}) + F_j(\text{SPT})}{n} \quad [5.17]$$

$$F_{\text{NON-SPT}} - F_{\text{SPT}} = \frac{(t+p_i) + (t+p_i+p_j) - (t+p_i+p_j) - (t+p_j)}{n} \quad [5.18]$$

$$F_{\text{NON-SPT}-F_{\text{SPT}}} = \frac{p_i - p_j}{n} > 0 \quad [5.19]$$

We then deduce that $F_{\text{NONSPT}} - F_{\text{SPT}} > 0$, and therefore to minimize the desired criterion, it is sufficient to follow the SPT method.

Let us consider the example in Table 5.7.

i	1	2	3	4
p_i	4	3	2	1

Table 5.7. *The data for the SPT example*

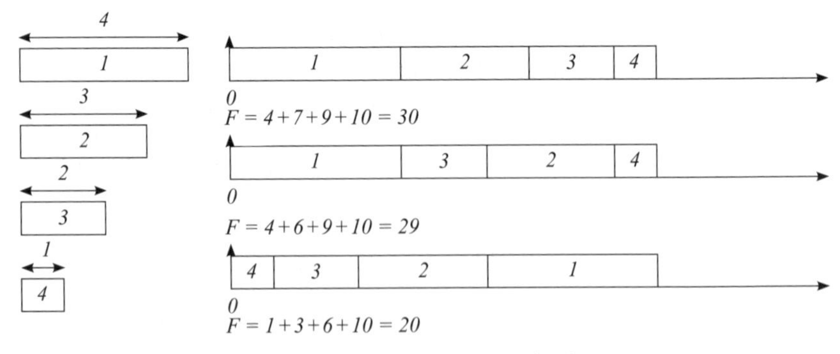

Figure 5.28. *An SPT example -1*

We may observe that the first two solutions in Figure 5.28 have a flow time greater than the time obtained per the SPT order in the third proposed solution, where 20 is the optimal solution to the problem.

5.2.5.2. *Minimization of the mean weighted flow time* $1/r_i = 0/\sum w_i C_i$

This is the same problem as before, but each job i has a weight, denoted by w_i, which allows its importance to be weighted relative to the others. The problem is therefore still polynomial, and may be resolved by an adaptation of the SPT method, called the weighted shortest processing time (WSPT), whose optimality is demonstrated on the same principle as for the SPT method. The jobs are arranged in increasing order of their durations divided by their weights.

5.2.5.3. *Minimization of the mean flow time* $1/r_i, pmtn/\sum C_i$

As considered earlier, the problem $1/r_i = 0/\sum C_i$ is polynomial. However, $1/r_i/\sum C_i$ is not. To find a lower bound (LB) for the NP-hard problem, one of the possible strategies is to relax one constraint. Implicitly, in $1/r_i/\sum C_i$, the jobs cannot be interrupted once they have begun. Now, by making them interruptible, i.e. by allowing pre-emption, the problem may be relaxed, becoming $1/r_i, pmtn/\sum C_i$, which is itself of polynomial time as shown in Baker's 1974 study [BAK 74].

The method used to resolve the problem is that of the shortest remaining processing time (SRPT). We therefore schedule jobs in increasing order of their remaining durations.

To prove the optimality of this method for the problem, we will consider two cases. The first case is that at time t, we have a single job pending. It is then necessary to show that any idle time introduced into the machine at this time will cause an increase in the value of the criterion $\sum C_i$ (Figure 5.29).

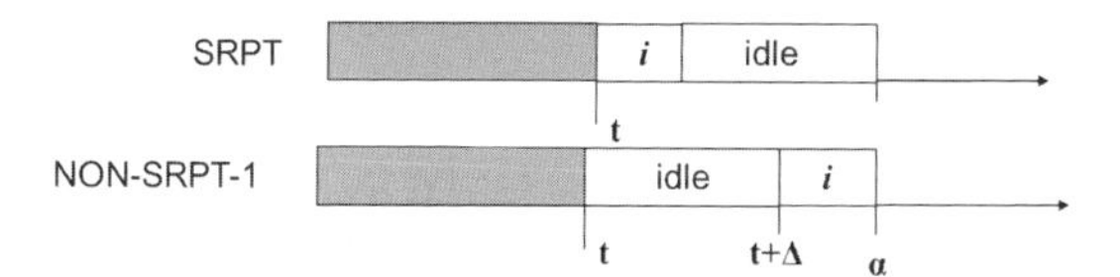

Figure 5.29. *An SRPT example -(1)*

It is obvious that in this case the removal of the idle time decreases the flow time and therefore the $\sum C_i$. Indeed, we have:

$$F_{NON-SRPT} - F_{SRPT} = \Delta/n > 0 \qquad [5.20]$$

The second case consists of comparing the placement of two jobs i and j whose remaining processing times $\tilde{p_i}$ and $\tilde{p_j}$ are such that $\tilde{p_i} < \tilde{p_j}$. Figure 5.30 compares an example following the SRPT method with a counter-example not following the SRPT rule.

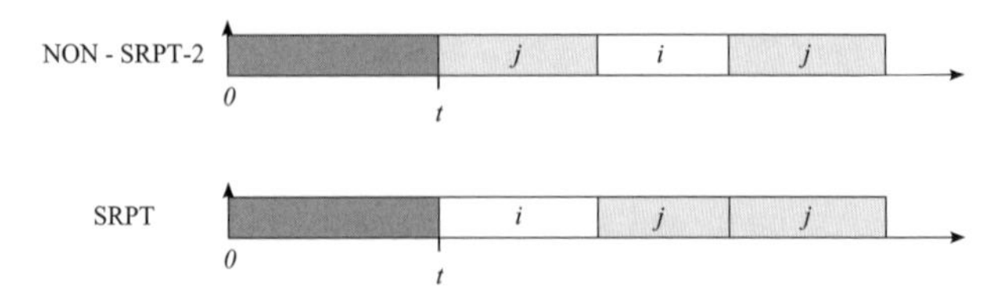

Figure 5.30. *An SRPT example -(2)*

EXAMPLE 5.3.– Consider the data in Table 5.8.

i	1	2	3	4
r_i	0	2	3	10
p_i	10	7	7	2

Table 5.8. *The data for the SRPT example*

At time $t = 0$, we schedule the only available job, job 1 (Figure 5.31).

Figure 5.31. *An SRPT example -(3)*

At time $t = 2$, job 2 arrives. Eight units of job 1 remain. It is therefore job 2 ($p_2 = 7$) that is scheduled until $t = 3$. job 3 arrives (Figure 5.32).

Figure 5.32. *An SRPT example -(4)*

However, at time $t = 3$, it is job 2 that has fewer remaining units (six). We therefore obtain the graph in Figure 5.33.

Figure 5.33. *An SRPT example -(5)*

At $t = 9$, job 2 is terminated. We have job 1 pending with eight units and job 3 with seven units. We thus schedule job 3 (Figure 5.34).

Figure 5.34. *An SRPT example -6*

At time $t = 10$, job 3 has six units remaining, job 1 has eight and job 4 has two. We therefore schedule job 4, then 3, and finally 1 (Figure 5.35).

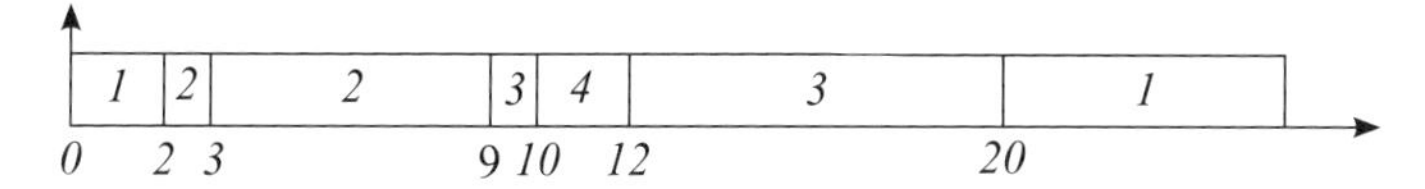

Figure 5.35. *An SRPT example -7*

5.2.5.4. *Minimization of the maximum tardiness* $T_{\max}$, $1/r_i = 0/T_{\max}$

This criterion corresponds to the minimization of the tardiness of the job with the greatest tardiness:

$$T_{\max} = \max_{i=1,\ldots,n} T_i \qquad [5.21]$$

This problem is polynomial and can be optimally resolved by the earliest due date (EDD) algorithm. In this algorithm the jobs are scheduled in increasing order of their due dates.

The proof of the optimality of this method can be validated on the same principle as for the previous one. Let us consider two schedules, written S (which does not follow the EDD rule) and S' (which does follow the EDD rule), which differ only in the permutation of two jobs i and j such that $d_i < d_j$, as shown in Figure 5.36.

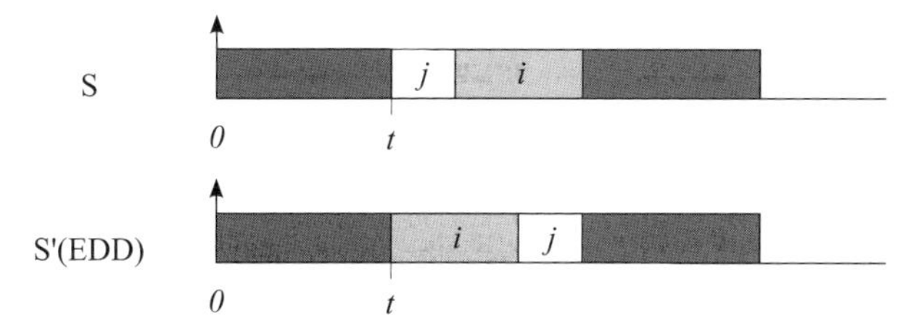

Figure 5.36. *The EDD proof*

We know by definition that:

$$T_i = \max(0, L_i) = \max(0, C_i - d_i) \qquad [5.22]$$

therefore:

$$T_{\max} = \max_{i=1,\ldots,n}(\max(0, C_i - d_i)) \qquad [5.23]$$

Let us first consider sequence S. We have:

$$T_i = \max(t + p_j + p_i - d_i, 0) \qquad [5.24]$$

and:

$$T_j = \max(t + p_j - d_j, 0) \qquad [5.25]$$

We then have:

$$\max(T_i, T_j) = \max(t + p_j + p_i - d_i, t + p_j - d_j, 0) \qquad [5.26]$$

We now consider sequence S' using the EDD method. We then have:

$$T'_i = \max(t + p_j - d_i, 0) \qquad [5.27]$$

and:

$$T'_j = \max(t + p_i + p_j - d_j, 0) \qquad [5.28]$$

We then have:

$$\max(T'_i, T'_j) = \max(t + p_j + p_i - d_j, t + p_i - d_i, 0) \qquad [5.29]$$

In conclusion, we have:

$$\max(T_i, T_j) = \max(t + p_j + p_i - d_i, t + p_j - d_j, 0) \qquad [5.30]$$

and:

$$\max(T'_i, T'_j) = \max(t + p_j + p_i - d_j, t + p_i - d_i, 0) \qquad [5.31]$$

Yet, by hypothesis, $d_j > d_i$, hence:

$$t + p_j + p_i - d_i > t + p_j - d_j \qquad [5.32]$$

$$t + p_j + p_i - d_i > t + p_j + p_i - d_j \qquad [5.33]$$

$$t + p_j + p_i - d_i > t + p_i - d_j \qquad [5.34]$$

$$t + p_j - d_j > t + p_j - d_i \qquad [5.35]$$

We thus deduce that:

$$\max(T_i, T_j) \geq \max(T'_i, T'_j) \quad [5.36]$$

Solution S', which follows the EDD sequence, is therefore better than solution S.

5.2.5.5. *Minimization of the maximum tardiness when the jobs have different arrival dates, with pre-emption* $1/r_i, pmtn/T_{\max}$

This problem was studied in 1976 by Blazewicz [BAZ 77]. The problem $1/r_i/T_{\max}$ is NP-hard, but by adding the possibility of the pre-emption of jobs, it becomes polynomial. The method used is called the Jackson sequence, and consists of giving priority to a new job arriving with a smaller delay. We interrupt the job in progress to carry out the new available job.

5.2.5.6. *Minimization of the mean tardiness* $1//\overline{T}$

In this problem, we implicitly assume that all of the jobs are available at the same time, therefore $r_i = 0$.

This problem is NP-hard, according to Lawler [LAW 77]. In certain scenarios, notably $1/p_i = p, r_i/\overline{T}$ and $1/pmtr, p_i = p, r_i/\overline{T}$, the problem is polynomial [BAP 00].

Regarding $1//\overline{T}$, numerous methods (approximate and exact) have been developed to solve this problem [CHU 92a].

5.2.5.6.1. Approximate method

In 1992, Chu and Portmann proposed a heuristic that gives an approximate solution to the problem [CHU 92c]. This method is based on a property that is true for every pair of consecutive jobs i and j.

Indeed, the authors showed that there exists an optimal solution such that for every pair of consecutive jobs i and j, the following relation is satisfied:

$$\max(t + p_i, d_i) \leq \max(t + p_j, d_j) \quad [5.37]$$

where t is the start date of job i in the schedule. We therefore schedule job i, and then job j.

Starting from this property, the authors proposed the heuristic approach presented by Algorithm 5.1 where L represents the list of jobs not yet scheduled and t the time variable.

Algorithm 5.1 The Chu and Portmann heuristic [CHU 92c]

1: L={1,2 ... n}
2: t=0
3: **while** $L \neq \emptyset$ **do**
4: $j^* = \arg\min_L \max(t + p_j, d_j)$
5: Schedule j^* at time t
6: $L = L - j^*$
7: $t = t + p_j$
8: **end while**

5.2.5.6.2. Exact method

To obtain an exact solution (but which is much more costly in calculation time), we may apply a branch and bound procedure. This uses an LB and dominance properties.

5.2.5.6.3. Lower bound

Consider a set of n jobs that we wish to schedule. Let $C_{[i]}$ be the end date of the job that terminates at position i in a schedule constructed using the SPT method. Let $d_{[i]}$ be the due date i.

The quantity in expression [5.38] is a LB, i.e. the optimum total delay cannot be less than this quantity [CHU 92c]. This calculation gives for each subset of jobs an optimistic evaluation of the delay:

$$\sum_{i=1}^{n} \max(C_{[i]} - d_{[i]}, 0) \quad [5.38]$$

Two dominance properties may be used.

5.2.5.6.4. Dominance property 1

There exists an optimal schedule such that if $p_i < p_j$ and $\max\{t + p_i, d_i\} \leq \max\{t + p_j, d_j\}$, then i precedes j, where t is the lower of the start dates of these two jobs.

The proof of this property draws on Figure 5.37. Sequence S does not satisfy the property, whereas S' does. We now calculate the tardiness of jobs i and j in the two cases and show that the sum $T_i + T_j$ (in the case that satisfies the dominance property) is smaller than $T'_i + T'_j$ (in the case that does not satisfy the dominance property):

$$T_j = \max(0; t + p_j - d_j) = \max(t + p_j; d_j) - d_j \quad [5.39]$$

$$T_i = \max(0; t + p_j + \Delta + p_i - d_i) = \max(t + p_j + \Delta + p_i; d_i) - d_i \quad [5.40]$$

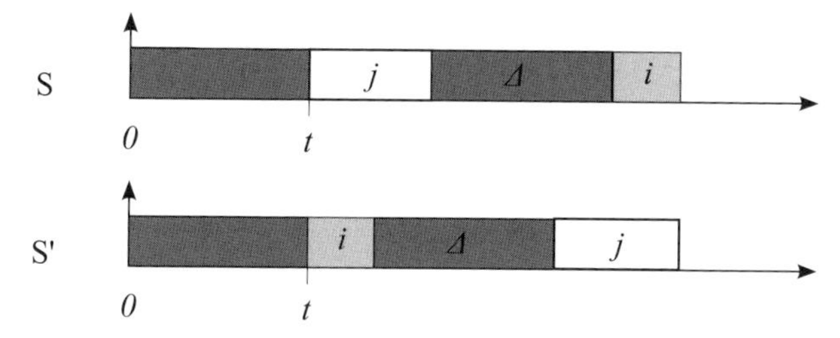

Figure 5.37. *Proof of dominance property 1*

We then show that:

$$T_j + T_i = \max(t + p_j + \Delta + p_i + \max(t + p_j; d_j); \max(t + p_j; d_j) + \max(t + p_i; d_i)) - (d_j + d_i) \quad [5.41]$$

However, we may similarly show that:

$$T'_j + T'_i = \max(t + p_j + \Delta + p_i + \max(+p_i; d_i); \max(t + p_i; d_i) + \max(t + p_j; d_j)) - (d_j + d_i) \quad [5.42]$$

then:

$$T'_j + T'_i < T_j + T_i \quad [5.43]$$

5.2.5.6.5. Dominance property 2

There exists an optimal schedule such that if $d_i > d_j$ and $d_i + p_i > t$, then job j precedes job i where t is the greater of the end dates of jobs i and j.

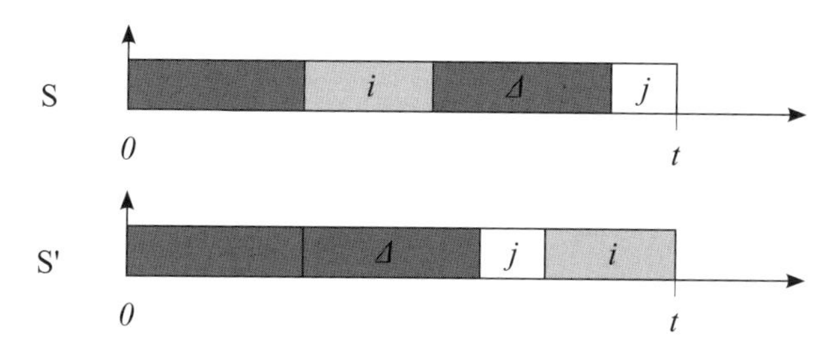

Figure 5.38. *Proof of dominance property 2*

The proof of this property draws on Figure 5.38. Sequence S does not satisfy dominance property 2, whereas S' does. We follow the proof of the previous property:

$$T_i = \max(0; C_i - d_i) = \max(0; t - p_j - \Delta - d_i) \quad [5.44]$$

$$T_j = \max(0; C_j - d_j) = \max(0; t - d_j) \quad [5.45]$$

We then have:

$$T_i + T_j = \max(2t - p_i - \Delta - d_i - d_j; t - p_j - \Delta - d_i; t - d_j; 0) \quad [5.46]$$

Similarly, we may show that:

$$T_i' + T_j' = \max(2t - p_i - d_i - d_j; t - p_i - d_j; t - d_i; 0) \quad [5.47]$$

Yet as $d_i > d_j$, therefore $t - d_i < t - d_j$ and $t - d_i - d_j < t - d_j$. We know that $d_i + p_i > t$ therefore $2t - d_i - p_i - d_j < t - d_j$. Then $T_i' + T_j' < T_i + T_j$.

EXAMPLE 5.4.– On day D, a company receives four work orders, referred to as projects 1, 2, 3 and 4. During discussions with the customers, the company manager identified the date at which each customer wishes to see his/her project finished. Of course, the company may propose later dates and negotiate, as the size of the company means that it is impossible to carry out more than one project at the same time. To simplify, we assume that the discount the company grants to its customers is proportional to the difference between the proposed project end date and the date desired by the customer (€500/day). The company wishes to minimize the financial loss generated by these negotiations. The estimated durations (in days) of the projects as well as the end dates desired by the customers (in number of working days from day D) are shown in Table 5.9.

Project	1	2	3	4
Estimated duration	17	22	25	28
Desired end date	72	68	40	45

Table 5.9. *Example 2 of parallel machines*

To obtain an upper bound (UB) for this problem, we may apply the EDD method. Following this order, the projects are executed in the order $3 - 4 - 2 - 1$ with end dates $C_3 = 25, C_4 = 53, C_2 = 75$ and $C_1 = 92$. Job 3 is not late, and the tardinesses of the other jobs are $T_4 = 8, T_2 = 7$ and $T_1 = 20$, i.e. $\sum T_i = 35$. We may therefore conclude that the optimal solution does not have a tardiness greater than 35.

Next, we choose to divide the set of solutions into four subsets: solutions that begin with job 1, those that begin with job 2, those that begin with job 3 and those that begin with job 4.

Let us examine subset 1. Job 1 is not late and finishes at time 17. To optimistically estimate of the quality of these solutions, we apply the calculation of the LB to the three remaining jobs, and obtain $LB_1 = 43$. The solutions in this subset therefore do not give solutions with a sum of tardinesses less than 43. Yet we already know one solution (given by EDD) with a total tardiness of 35. It is therefore not necessary to develop the solutions of this subset.

Regarding subset 2, the LB is 39. We may therefore also eliminate this solution subset. For the solution subset from job 3, we have $LB_3 = 20$ and for subset 4 we have $LB_4 = 25$. It is therefore necessary to develop these two subsets.

Let us begin by developing node 3. If we schedule job 1 after job 3, the LB gives an estimate of 43, and in the case where we schedule job 2 after job 3, the estimate is 39. We will therefore not examine these nodes any further as the estimate of the LB is greater than the best known solution that has a tardiness of 35. The LB is 30 when job 4 is scheduled after job 3. We, therefore, examine the two possible solutions of this subset: the solution $3 - 4 - 1 - 2$, which has a total tardiness of 32, and the solution $3 - 4 - 2 - 1$, which has a total tardiness of 35. The new best known solution is now $3 - 4 - 1 - 2$, with a tardiness of 32. The solution subset from node 4 now remains to be examined. When we schedule job 3 after job 4, the LB gives a tardiness of at least 35, which is greater than 32. As for the two other subsets coming from node 4, they are also dominated. The optimal solution is therefore to schedule the jobs in the order $3 - 4 - 1 - 2$ for a tardiness of 32.

If we also wished to use dominance properties, we would be able to note that it is useless to study the solutions where job 1 is before job 3. Indeed, for $t = 0$, we have $\max\{0 + 25; 40\} < \max\{0 + 17; 72\}$.

5.2.5.7. *Minimization of the flow time* $1/r_i/\overline{F}$

This problem is NP-hard (Lenstra *et al.* [LEN 77]). We know that when all of the jobs are available at time $t = 0$, the problem is of polynomial time, and may be solved using the SPT heuristic. This may give interesting solutions for the $1/r_i/\overline{F}$ problem, but is not optimal.

Some studies show that the filing of jobs in the FIFO order also sometimes gives interesting solutions. The PRTF method is inspired by this observation [CHU 92b].

5.2.5.7.1. $PRTF$ method

The PRTF function is a dynamic measure of priority: the smaller the value of the function, the greater the priority of the corresponding job.

The $PRTF(i, \Delta)$ function, for a job i at time Δ, is defined as follows:

$$PRTF(i, \Delta) = 2\max(r_i, \Delta) + p_i \quad [5.48]$$

If there are only two jobs to be scheduled on one machine from time Δ, the necessary and sufficient condition for scheduling job i before job j is that:

$$PRTF(i, \Delta) \leq PRTF(j, \Delta) \quad [5.49]$$

5.2.5.7.2. EST method

The earliest starting time (EST) algorithm consists of scheduling the jobs as soon as possible, that is in the order of the r_i, giving priority to the shortest jobs when several jobs are eligible.

5.2.5.7.3. $APRTF$ method

At a given time t when the machine is available, we write α for the job to be scheduled according to the PRTF method and β for the job to be scheduled according to the EST method (Figure 5.39). Hence, there are two possible options:

– if $\alpha = \beta$, there is no choice to be made, and we schedule this job;

– if $\alpha \neq \beta$, a choice must be made.

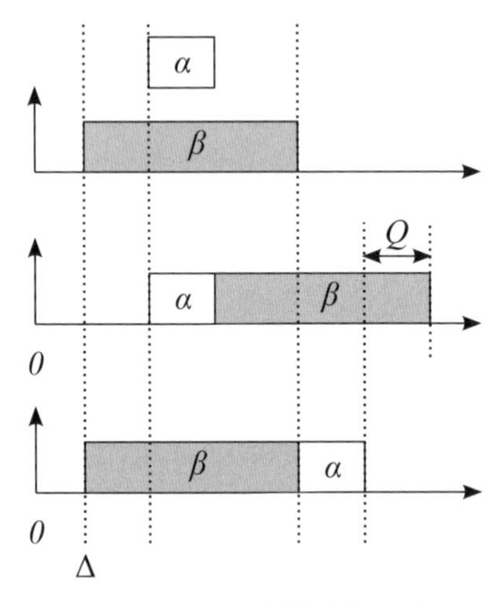

Figure 5.39. *The $APRTF$ method*

It is then necessary to calculate the gain G for the flow time between the two possible operations (schedule α then β or β then α):

$$G = F_{\beta,\alpha}(\Delta) - F_{\alpha,\beta}(\Delta) \quad [5.50]$$

We write μ for the number of jobs remaining to be scheduled. We then compare G with the product $\mu \times Q$, and we have:

– if $G > \mu \times Q$, then we schedule job α;

– if $G < \mu \times Q$, then we schedule job β.

5.2.6. *Scheduling a flow shop workshop*

Flow shop-type workshops are encountered when a linear manufacturing process comprises a set of machines visited in the same order by all of the products manufactured by this set. The machines are therefore arranged in the order in which they are visited.

However, the processing time on each machine may differ for different products.

For this type of production system, the objectives are many and varied. They may be to reduce the processing time of products in the system, or to respect delays. However, there is one very common objective, which is to minimize the makespan ($C_{\max}$), that is the difference between the end time of the manufacture of the last issued product and the start time of the manufacture of the first issued product. The latter depends on the order in which the products are issued.

In the case of two machines, this problem is of polynomial time. If there are more than two machines, the problem becomes NP-hard.

The following constraints must be respected:

– all of the jobs must occur on all of the machines;

– a machine may treat only a single job at once.

To solve this problem, we must study the permutation schedules. Such schedules are such that if two products p_1 and p_2 cross a flow shop, p_1 goes before p_2 on the first machine, which implies that the same order is followed on all of the other machines.

5.2.6.1. *The two-machine problem*

In the case where the flow shop problem involves two machines, it may be optimally resolved with the Johnson algorithm [JOH 54]. The Johnson rule minimizes the delay for obtaining several orders by arranging the order of batch scheduling.

5.2.6.1.1. The Johnson rule

Job i precedes job j in the optimal sequence of the two-machine flow shop if:

$$\min(p_{i,1}, p_{j,2}) < \min(p_{i,2}, p_{j,1}) \qquad [5.51]$$

where $p_{i,1}$ is the operational time of job i on machine 1, and $p_{i,2}$ that on machine 2.

i	1	2	3	4	5
$p_{i,1}$	4	1	9	3	9
$p_{i,2}$	2	6	6	8	4

Table 5.10. *An example of the Johnson rule*

5.2.6.1.2. Implementation

We consider two lists denoted $Sstart$ and $Send$. The jobs i are distributed such that $p_{i,1} < p_{i,2}$ in $Sstart$, and $p_{i,1} > p_{i,2}$ in $Send$. The jobs that are in $Sstart$ must be arranged in increasing order of the $p_{i,1}$, and the jobs that are in $Send$ must be arranged in decreasing order of the $p_{i,2}$. The jobs in $Sstart$ must be executed before those in $Send$.

5.2.6.1.3. An example of application

In $Sstart$, we get jobs 2, 4 and in $Send$, jobs $1, 3, 5$. Finally, we obtain the optimal sequence $2 - 4 - 3 - 5 - 1$.

5.2.6.2. *A particular case of the three-machine problem*

Generally, the problem is NP-hard as soon as there are more than two machines. We may, however, find the optimal solution in certain three-machine scenarios. These are the cases where machine 2, which is in the middle, is dominated by the others, i.e. if:

– the time on machine 1 is greater than on machine 2:

$$\min_k p_{k,1} \geq \max_k p_{k,2}$$

In this case, job i must be placed before job j in the optimal sequence if:

$$\min_k p_{i,1} + p_{i,2}, p_{j,2} + p_{j,3} \leq \min_k p_{i,2} + p_{i,3}, p_{j,1} + p_{j,2}$$

– the time on machine 3 is greater than on machine 2:

$$\min_k p_{k,3} \geq \max_k p_{k,2}$$

In this case, job i must be placed before job j in the optimal sequence if:

$$\min_k p_{i,1} + p_{i,2}, p_{j,2} + p_{j,3} \leq \min_k p_{i,2} + p_{i,3}, p_{j,1} + p_{j,2}$$

5.2.6.3. *The m-machine problem*

m-machine flow shop problems may be resolved by heuristics such as those of Palmer, Gupta and Campbell, Dudek and Smith.

5.2.6.3.1. The Palmer heuristic [PAL 65]

In 1965, Palmer [PAL 65] proposed a decreasing index ranking for scheduling jobs on machines as a function of execution times. The idea is to give priority to the jobs with execution times that increase during the work schedule.

This heuristic is inspired by the Johnson rule which says that a product with a range start smaller than the end of the range goes in front of a product with a greater range start.

This heuristic is slightly different from the Johnson rule as it does not guarantee optimality, but can be applied to problems with two or more machines.

For this, we calculate the value of the index s_j:

$$s_j = \sum_{k=1}^{m} (2k - m - 1) p_{j,k} \quad [5.52]$$

The sequence is determined by ranking the jobs in decreasing order of their indices.

The smaller the start of the schedule and the greater the end of the schedule, the greater the s_j.

5.2.6.3.2. The Gupta heuristic [GUP 71]

In 1971, Gupta [GUP 71] presented another heuristic, very similar to that of Palmer, except in the calculation of the index. The index s_j is calculated using the following formula:

$$s_j = \frac{e_j}{\min_{1 \leq k \leq m-1} (p_{j,k} + p_{j,k+1})} \quad [5.53]$$

$$e_j = \begin{cases} 1 & \text{if } p_{j,k} < p_{j,m} \\ -1 & \text{otherwise} \end{cases} \quad [5.54]$$

The sequence is determined by ranking jobs in decreasing order of their indices.

5.2.6.3.3. The Campbell, Dudek and Smith heuristic [CAM 71]

In 1970, Campbell, Dudek and Smith [CAM 71] proposed a method generalizing the Johnson algorithm. The efficiency of this method lies in its use of the Johnson rule and its creation of several schedules from which the best order is chosen. This heuristic is better known by the acronym CDS, and its performance is generally better than that of the Palmer heuristic. The heuristic is summarized in Algorithm 5.2.

Algorithm 5.2 The CDS heuristic [CAM 71]

1: **for** $j = 1$ to $m - 1$ **do**
2: Consider the problem with two fictitious machines $(M1, M2)$, where:
3: the duration on $M1$ is the sum of the durations on machines 1 to j
4: the duration on $M2$ is the sum of the durations on machines $m + 1 - j$ to m
5: Apply the Johnson rule to $M1$ and $M2$
6: Hence deduce an order for the jobs
7: Calculate the total duration for the initial problem
8: **end for**

5.2.7. *Parallel-machine problems*

In the context of problems with parallel machines, each job involves one operation and we make use of m machines. This job may be executed on one of the machines.

These m machines may be:

– identical: the operational time of a job is the same on all of the machines;

– uniform: the performances of the machines are all proportional: there is a coefficient α_k for each machine k $(\neq 1)$ such that the operational time of a job on machine k is α_k times that of machine 1; and

– unrelated: there is no general relation between the performances of the machines.

We also distinguish the case where a job may be executed only on a single machine from the case where it is possible to divide a job into several sub-jobs carried out by several machines. In the second case, it is necessary to distinguish the case with pre-emption, where there can be no superposition (Figure 5.40), from the case of splitting, where superposition is authorized (Figure 5.41).

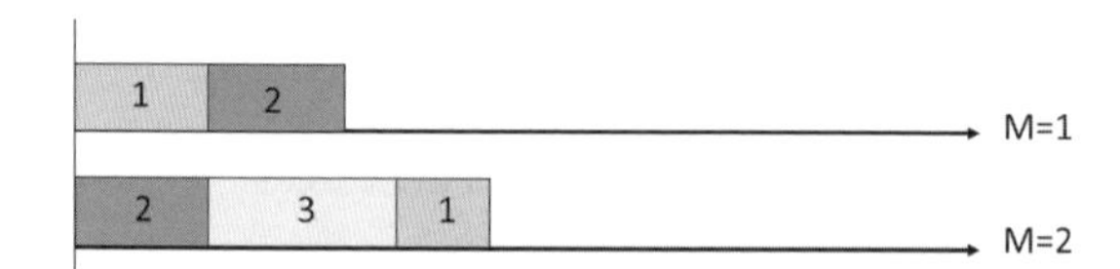

Figure 5.40. *Parallel machines, pre-emption*

5.2.7.1. *Identical machines, $r_i = 0$, $Min\ F$*

In this problem, we consider all of the jobs to be available at time $t = 0$. The problem is of polynomial time. The resolution method consists of ranking the jobs in non-decreasing order of their operational time p_i. As soon as a machine is available, the job at the top of the list is allocated to it.

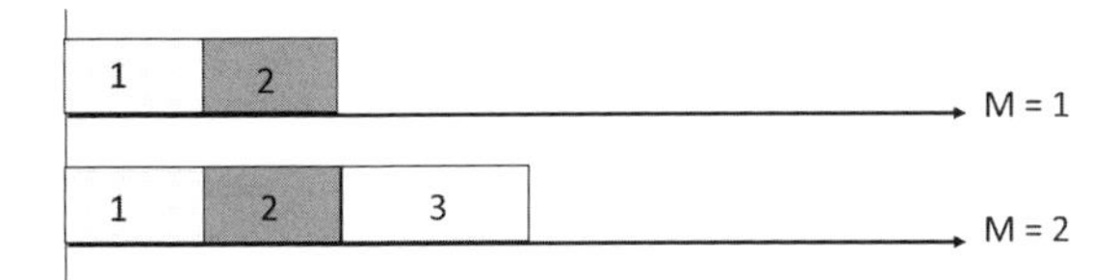

Figure 5.41. *Parallel machines, splitting*

To minimize the flow time, we apply here the SPT method, but over several machines. Yalaoui and Chu show [YAL 06] that if the machines have different availability dates, the problem remains of polynomial time. They then generalize this method, calling it the generalized SPT method (SPTG).

In the case where all of the jobs are not available at $t = 0$, the problem is then NP-hard, even if pre-emption is authorized. Some exact methods, based on dynamic programming, branch and bound procedures and LBs, as well as approximate methods, have been developed.

5.2.7.2. *Identical machines, $r_i = 0$, $Min\ C_{\max}$, interruptible jobs*

In this problem, we still have m identical machines and the jobs are interruptible, and all available at time 0.

It is fairly simple to obtain in this case an LB with the following formula:

$$LB = \max \left\{ \frac{1}{m} \sum_{i=1}^{n} p_i ; \max_{1 \leq i \leq n} p_i \right\} \quad [5.55]$$

If we are capable of obtaining a solution that achieves this bound, then it is the optimal solution. For this, it is sufficient to successively fill, up to the maximum, the machines until the value of LB is attained. The allocation of a job is interrupted when LB is attained.

Bibliography

[ABD 06] ABDUL-KADER W., "Capacity improvement of an unreliable production line: an analytical approach", *Computers and Operations Research*, vol. 33, pp. 1695–1712, 2006.

[ALT 02] ALTIPARMAK F., BUGAK A., DENGIZ B., "Optimization of buffer sizes in assembly systems using intelligent techniques", *Proceedings of the 34th Winter Simulation Conference*, San Diego, CA, pp. 1157–1162, 2002.

[ALV 00] ALVARENGA A.D., NEGREIROS-GOMES F., MESTRIA M., "Metaheuristic methods for a class of the facility layout problem", *Journal of Intelligent Manufacturing*, vol. 11, pp. 421–430, 2000.

[AMO 99] AMODEO L., Contribution à la simplification et à la commande des réseaux de Petri stochastiques. Application aux systèmes de production, Thesis, University of Franche-Comté, Belfort, 1999.

[AMO 05] AMODEO L., YALAOUI F., *Logistique interne: entreposage et manutention*, Ellipses collection TechnoSup, Paris, 2005.

[AMO 07] AMODEO L., CHEN H., HADJI A.E., "Multi-objective supply chain optimization: an industrial case study", *Lecture Notes in Computer Science*, vol. 4448/2007, pp. 732–741, 2007.

[ARC 66] ARCUS A., "COMSOAL: a computer method of sequencing operations for assembly lines", *International Journal of Production Research*, vol. 4, pp. 259–277, 1966.

[AUL 72] AULEY A.M., "Machine grouping for efficient production", *Production Engineer*, vol. 51, pp. 53–57, 1972.

[BAI 80] BAILLARGEON G., *Introduction aux méthodes statistiques en contrôle de la qualité*, Les Editions SMG, Trois-Rivières, pp. 151, 1980.

[BAK 74] BAKER K., *Introduction to Sequencing and Scheduling*, John Wiley, New York, 1974.

[BAP 00] BAPTISTE P., "Scheduling equal-length jobs on ientical paralle machines", *Discrete Applied Mathematics*, vol. 103, pp. 21–32, 2000.

[BAY 86] BAYBARS I., "A survey of exact algorithms for the simple assembly line balancing problem", *Management Science*, vol. 32, pp. 909–932, 1986.

[BAZ 77] BAZEWICZ J., "Scheduling dependent tasks with different arrival times to meet deadlines", *Proceedings of the International Workshop organized by the Commision of the European Communities on Modelling and Performance Evaluation of Computer Systems*, Stresa, North-Holland Publishing Co., Amsterdam, The Netherlands, pp. 57–65, 1977.

[BEL 57] BELLMAN R., *Dynamic Programming*, Princeton University Press, Princeton, 1957.

[BEN 07] BENLIAN X., ZHIQUAN W., "A multi-objective-ACO-based data association method for bearings-only multi-target tracking", *Communications in Nonlinear Science and Numerical Simulation*, vol. 12, no. 8, pp. 1360–1369, 2007.

[BIT 82] BITRANE G., YANASSE H., "Computational complexity of the capacitated lot size problem", *Management Science*, vol. 28, no. 10, pp. 1174–1186, 1982.

[BOU 03] BOUTEVIN C., GOURGAND M., NORRE S., "Méthodes d'optimisation pour le problème de l'équilibrage de lignes d'assemblage", *Proceedings of the MOSIM'2003 Conference*, Toulouse, France, 2003.

[BOY 06] BOYSEN N., FLIEDNER M., SCHOLL A., "Assembly line balancing: which model to use when?" *International Journal of Production Economics*, vol. 111, no. 2, pp. 509–528, 2006.

[BRA 06] BRAHIMI N., DAUZERE-PERES S., NAJID M., NORDLI A., "Single item lot sizing problems", *European Journal of Operational Research*, vol. 168, pp. 1–16, 2006.

[BRO 06] BROWN E., SUMICHRAST R., "Evaluating performance advantages of grouping genetic algorithms", *Engineering Applications of Artificial Intelligence*, vol. 18, pp. 1–12, 2006.

[BUK 00] BUKCHIN J., TZUR M., "Design of flexible assembly line to minimize equipment cost", *IIE Transactions*, vol. 32, no. 7, pp. 585–598, 2000.

[BUK 03] BUKCHIN J., RUBINOVITZ J., "A weighted approach for the assembly line design with station paralleling and equipment selection", *IIE Transactions*, vol. 35, pp. 73–85, 2003.

[BUR 97] BURMAN M., GERSHWIN S., Analysis of continuous material models of unreliable flow lines, Report, Massachsetts Institute of Technology, Massachusetts, MA, 1997.

[BUZ 71] BUZACOTT J., "The role of inventory banks in flow-line production systems", *International Journal of Production Research*, vol. 9, no. 4, pp. 425–436, 1971.

[CAM 71] CAMPBELL H., DUDEK R., SMITH M., "A heuristic algorithm for the n job m machine sequencing problem", *Management Science B*, vol. 16, pp. 630–637, 1971.

[CET 07] CETINKAYA K., "Design and application an integrated element selection model for press automation line", *Materials and Design*, vol. 28, pp. 217–229, 2007.

[CHE 77] CHEVALIER A., *La programmation dynamique*, Dunod, Paris, 1977.

[CHE 92] CHERN M., "On the computational complexity of reliability redundancy allocation in a series system", *Operations Research Letters*, vol. 11, pp. 309–315, 1992.

[CHE 07] CHEHADE H., AMODEO L., YALAOUI F., GUGLIELMO P.D., "Optimisation multiobjectif appliquée au problème de dimensionnement de buffers", *Proceedings of LT'07 Logistique et Transport*, Sousse, Tunisia, pp. 337–342, November 2007.

[CHE 08a] CHEHADE H., YALAOUI F., AMODEO L., "A simulation-based optimization method applied to a printing workshop", *Advances in Management*, vol. 1, no. 3, pp. 42–49, 2008.

[CHE 08b] CHEHADE H., YALAOUI F., AMODEO L., GUGLIELMO P.D., "Ant colony optimization for assembly lines design problem", *Proceedings of the 8th International Conference on Computational Intelligence FLINS08*, Madrid, Spain, pp. 1135–1140, 2008.

[CHE 09] CHEHADE H., YALAOUI F., AMODEO L., GUGLIELMO P.D., "Optimisation multiobjectif pour le problème de dimensionnement de buffers", *Journal of Decision Systems*, vol. 18, no. 2, pp. 257–287, 2009.

[CHE 11] CHEHADE H., AMODEO L., YALAOUI F., "Ant colony optimization for multiobjective buffers sizing problems", *Ant Colony Optimization - Methods and Applications, Avi Ostfled, InTech*, Chapter 19, pp. 303–316, 2011.

[CHO 87] CHOW W., "Buffer capacity analysis for sequential production line with variable process times", *International Journal of Production Research*, vol. 25, no. 8, pp. 1183–1196, 1987.

[CHU 92a] CHU C., "A branch and bound algorithm to minimize total tardiness with different release dates", *Naval Research Logistics*, vol. 39, pp. 265–283, 1992.

[CHU 92b] CHU C., "Efficient heuristics to minimize total flow time with release dates", *Original Research Article Operations Research Letters*, vol. 12, pp. 321–330, 1992.

[CHU 92c] CHU C., PORTMANN M., "Some new efficient methods to solve the n/1/ri/[epsilon]Ti scheduling problem", *European Journal of Operational Research*, vol. 58, pp. 404–413, 1992.

[COE 07] COELLO C., AGUIRRE A., ZITZLER E., "Evolutionary multi-objective optimization", *European Journal of Operational Research*, vol. 181, no. 16, pp. 1617–1619, 2007.

[COL 02] COLLETTE Y., SIARRY P., *Optimisation multiobjectif*, Eyrolles, Paris, 2002.

[DAN 06] DANIEL J., RAJENDRAN C., "Heuristic approaches to determine base stock levels in a serial supply chain with a single objective and with multiple objectives", *European Journal of Operational Research*, vol. 175, pp. 566–592, 2006.

[DAV 89] DAVID R., ALLA H., *Du grafcet au réseau de Petri*, Hermès, traités des nouvelles technologies, 1989.

[DEB 94] DEB K., SRINIVAS N., "Multiobjective optimization using non-dominated sorting in genetic algorithms", *Evolutionary Computation*, vol. 2, no. 3, pp. 221–248, 1994.

[DEB 02] DEB K., PRATAP A., AGARWAL S., MEYARIVAN T., "A fast and elitist multi-objective genetic algorithm: NSGA - II", *IEEE Transactions on Evolutionary Computation*, vol. 6, no. 2, pp. 182–197, 2002.

[DES 05] DESCOTES-GENON B., "Problème de logistique inverse: utilisation d'une méta-heuristique dans une application de transfert de marchandises", *Revue REE*, vol. 1, pp. 34–40, 2005.

[DHA 10] DHAENENS C., LEMESRE J., TALBI E., "K-PPM: a new exact method to solve multi-objective combinatorial optimization problems", *European Journal of Operational Research*, vol. 200, pp. 45–53, 2010.

[DOL 02] DOLGUI A., EREMEEV A., KOLOKOLOV A., SIGAEV V., "A genetic algorithm for the allocation of buffer storage capacities in a production line with unreliable machines", *Journal of Mathematical Modelling and Algorithms*, vol. 1, pp. 89–104, 2002.

[DOL 03] DOLGUI A., MAKDESSIAN L., YALAOUI F., "Equipment cost minimization for transfer lines with blocks of parallel operations", *Proceedings of the IEPM'03 Conference*, Porto, Portugal, pp. 412–421, 2003.

[DOL 06a] DOLGUI A., PROTH J., *Les systèmes de production modernes. Tome 1: Conception, gestion et optimisation*, Hermès, Paris, 2006.

[DOL 06b] DOLGUI A., PROTH J., *Les systèmes de production modernes. Tome 2: Outils et corrigés des exercices*, Hermès, Paris, 2006.

[DOR 92] DORIGO M., Optimization, learning and natural algorithms, PhD Thesis, Politecnico di Milano, 1992.

[DRA 96] DRAKE G., SMITH J., "Simulation system for real-time planning, scheduling and control", *Proceedings of the 28th Winter Simulation Conference*, Coronado, CA, pp. 1083–1090, 1996.

[DRE 05] DREO J., PETROWSKI A., SIARRY P., TAILLARD E., *Métaheuristiques pour l'optimisation difficile*, Eyrolles, Paris, 2005.

[DSO 97] D'SOUZA K., KHATOR S., "System reconfiguration to avoid deadlocks in automated manufacturing systems", *Computers and Industrial Engineering*, vol. 32, no. 2, pp. 455–465, 1997.

[DUG 07] DUGARDIN F., AMODEO L., YALAOUI F., "Application de la dominance de Lorenz à l'ordonnancement multi-objectif", *Logistique et Transport 07 (LT'07)*, 2007.

[DUG 08] DUGARDIN F., YALAOUI F., AMODEO L., "Reentrant lines scheduling and Lorenz dominance: a comparative study", *Proceedings of the 8th International Conference on Computational Intelligence FLINS08*, Madrid, Spain, 2008.

[DUR 98] DURET D., M.PILLET, *Qualité en production*, Editions d'Organisation, Paris, 1998.

[DYC 90] DYCKHOFF H., "A typology of cutting and packing problems", *European Journal of Operational Research*, vol. 44, pp. 154–159, 1990.

[ELE 03] ELEGBEDE A., CHU C., ADJALLAH K., YALAOUI F., "Reliability allocation through cost minimization", *IEEE Transactions on Reliability*, vol. 52, pp. 106–111, 2003.

[ELM 98] EL-MARAGHY H., ABDALLAH I., EL-MARAGHY W., "On-line simulation and control in manufacturing systems", *Annals of the CIRP*, vol. 47, pp. 401–404, 1998.

[FEL 68] FELLER W., *An Introduction to Probability Theory and its Applications*, vol. 1, 3rd ed., John Wiley, New York, 1968.

[FLE 06] FLEURY G., LACOMME P., TANGUY A., *Simulation à événements discrets: Modèles déterministes et stochastiques*, Eyrolles, Paris, 2006.

[FLO 71] FLORIAN M., KLEIN M., "Deterministic production planning with concave with concave costs and capacity constraints", *Management Science*, vol. 18, pp. 12–22, 1971.

[FON 95] FONSECA C., FLEMING P., "An overview of evolutionary algorithms in multiobjective optimization", *Evolutionary Computation*, vol. 3, no. 1, pp. 1–16, 1995.

[FU 02] FU M., "Optimization for simulation: theory vs. practice", *Informs Journal on Computing*, vol. 14, pp. 192–215, 2002.

[GAG 02] GAGNÉ C., GRAVEL M., "Algorithme d'optimisation par colonie de fourmis avec matrices de visibilité multiples pour la résolution d'un problème d'ordonnancement industriel", *Infor*, vol. 40, no. 3, pp. 259–276, 2002.

[GAL 10] GALTIER J., LAUGIER A., http:perso.rd.francetelecom.fr/galtier/courses, 2010.

[GAR 79] GAREY M.R., JOHNSON D.S., *Computers and Intractability: A Guide to the Theory of NP Completeness*, Freeman, San Francisco, 1979.

[GER 87] GERSHWIN S., "An efficient decomposition method for the approximate evaluation of tandem queues with finite storage space and blocking", *Operations Research*, vol. 35, no. 2, pp. 291–305, 1987.

[GER 94] GERSHWIN S., *Manufacturing Systems Engineering*, Prentice Hall, Englewood Cliffs, 1994.

[GER 00] GERSHWIN S., SCHOR J., "Efficient algorithms for buffer space allocation", *Annals of Operations Research*, vol. 93, pp. 117–144, 2000.

[GHO 89] GHOSH S., GAGNON R., "A comprehensive literature review and analysis of the design, balancing and scheduling of assembly systems", *International Journal of Production Research*, vol. 27, no. 4, pp. 637–670, 1989.

[GLO 90] GLOVER F., "Tabu search: a tutorial", *Interfaces*, vol. 20, no. 4, pp. 74–94, 1990.

[GLO 96] GLOVER F., KELLY J., LAGUNA M., "New advances and applications of combining simulation and optimization", *Proceedings of the 28th Winter Simulation Conference*, Coronado, CA, pp. 144–152, 1996.

[GOL 85] GOLDBERG D.E., "Genetic algorithms and rule learning in dynamic system control", *Proceedings of the International Conference on Genetic Algorithms and Their Applications*, Pittsburgh, PA, pp. 8–15, 1985.

[GOL 89] GOLDBERG D., *Genetic Algorithms in Search, Optimization and Machine Learning*, Kluwer Academic Publishers, Boston, 1989.

[GRA 79a] GRAHAM R., LAWLER E., LENSTRA J., KAN A.R., "Optimisation and approximation in deterministic sequencing and scheduling", *Annals of Discrete Mathematics*, vol. 4, pp. 287–326, 1979.

[GRA 79b] GRAVES S., WHITNEY D., "A mathematical programming procedure for equipment selection and system evaluation in programmable assembly", *Proceedings of the 18th IEEE Conference on Decision and Control*, Fort Lauderdale, FL, pp. 531–536, 1979.

[GUP 71] GUPTA J., "A functional heuristic algorithm for the flow-shop scheduling problem", *Operational Research Quarterly*, vol. 22, pp. 39–47, 1971.

[HAI 71] HAIMES Y., LADSON L., WISMER D., "On a bicriterion formulation of the problems of integrated system identification and system optimization", *IEEE Transaction on System*, vol. 1, pp. 296–297, 1971.

[HAM 06] HAMADA M., MARTZ H., BERG E., KOEHLER A., "Optimizing the product-based availability of a buffered industrial process", *Reliability Engineering and System Safety*, vol. 91, pp. 1039–1048, 2006.

[HAN 06a] HANI Y., Optimisation physique et logique d'un établissement industriel: étude de l'EIMM de Romilly Sur Seine - SNCF, Thesis, University of Technology of Troyes, Troyes, 2006.

[HAN 06b] HANI Y., AMODEO L., YALAOUI F., CHEN H., "A hybrid genetic algorithm for solving an industrial layout problem", *Journal of Operations and Logistics*, vol. 1, no. 1, pp. IV.1–IV.17, 2006.

[HAN 06c] HANI Y., CHEHADE H., AMODEO L., YALAOUI F., "Simulation-based optimization of a train maintenance facility model using genetic algorithms", *Proceedings of the IEEE International Conference on Service Systems and Service Management*, Troyes, France, pp. 513–518, 2006.

[HAN 07] HANI Y., AMODEO L., YALAOUI F., CHEN H., "Ant colony optimization for solving an industrial layout problem", *European Journal of Operational Research*, vol. 183, pp. 633–642, 2007.

[HAR 95] HARMONOSKY C., "Simulation-based real-time scheduling: review of recent developments", *Proceedings of the 27th Winter Simulation Conference*, Arlington, VA, pp. 220–225, 1995.

[HAR 99] HARRIS J., POWELL S., "An algorithm for optimal buffer placement in reliable serial lines", *IIE Transactions*, vol. 31, pp. 287–302, 1999.

[HEN 92] HENDRICKS K., "The output processes of serial production lines of exponential machines with finite buffers", *Operations Research*, vol. 40, no. 6, pp. 1139–1147, 1992.

[HOF 92] HOFFMANN T., "EUREKA: a hybrid system for assembly line balancing", *Management Science*, vol. 38, no. 1, pp. 39–47, 1992.

[ISH 90] ISHIKAWA K., *Introduction to Quality Control*, Loftus 3A Corporation, Tokyo, 1990.

[ISH 03] ISHIBUCHI H., YOSHIDA T., MURATA T., "Balance between genetic search and local search in memetic algorithms for multiobjective permutation flowshop scheduling", *IEEE Transactions on Evolutionary Computation*, vol. 7, no. 2, pp. 204–223, 2003.

[JAF 89] JAFARI M., SHANTHIKUMAR J., "Determination of optimal buffer storage capacities and optimal allocation in multistage automatic transfer lines", *IIE Transactions*, vol. 21, no. 2, pp. 130–135, 1989.

[JOH 54] JOHNSON S., "Optimal two- and three-stage production schedules with setup times included", *Naval Research Logistics Quarterly*, vol. 1, pp. 61–68, 1954.

[JOH 88] JOHNSON R., "Optimally balancing large assembly lines with FABLE", *Management Science*, vol. 34, no. 2, pp. 240–253, 1988.

[KAC 01] KACEM I., HAMMADI S., BORNE P., "Direct chromosome representation and advanced genetic operators for flexible job shop problems", *Proceedings of the International Conference on Computational Intelligence for Modelling Control and Automation, CIMA 01*, Las Vegas, NV, 2001.

[KEN 95] KENNEDY J., EBERHART R., "Particle swarm optimization", *International Conference on Neural Networks*, pp. 1942–1948, 1995.

[KIL 61] KILBRIDGE M., WESTER L., "A heuristic method of assembly line balancing", *Journal of Industrial Engineering*, vol. XII, no. 4, pp. 292–298, 1961.

[KIR 83] KIRKPATRICK S., GELATT C., VECCHI M., "Optimisation by simulated anneling", *Science*, vol. 220, pp. 671–680, 1983.

[KOB 95] KOBAYASHI S., ONO I., YAMAMURA M., "An efficient genetic algorithm for job shop scheduling problems", *Proceeding of the 6th International Conference on Genetic Algorithms*, pp. 506–511, 1995.

[KOO 57] KOOPMANS T., BECKMAN M., "Assignment problems and the location of economic activities", *Econometrica*, vol. 25, pp. 53–76, 1957.

[KOR 99] KOREN Y., HEISEL U., JOVANE F., MORIWAKI T., PRITCHOW G., BRUSSEL H.V., "Reconfigurable manufacturing systems", *CIRP Annals*, vol. 48, no. 2, pp. 527–598, 1999.

[KUO 01] KUO W., PRASAD V., TILLMAN F., HWANG C., *Optimal Reliability Design: Fundamentals and Applications*, Cambridge, UK, 2001.

[LAC 03] LACOMME P., PRINS C., SEVAUX M., "Multiobjective capacitated arc routing problem", *Lecture Notes in Computer Science*, vol. 2632, pp. 550–564, 2003.

[LAC 06] LACOMME P., PRINS C., SEVAUX M., "A genetic algorithm for a bi-objective capacitated arc routing problem", *Computers and Operations Research*, vol. 33, no. 12, pp. 3473–3493, 2006.

[LAM 89] LAMOUILLE J.-L., MURRY B., POTIÉ C., *La Maîtrise Statistique des Procédés*, AFNOR Gestion, Paris, 1989.

[LAU 00] LAUMANNS M., ZITZLER E., THIELE L., "A unified model for multi-objective evolutionary algorithms with elitism", *Congress on Evolutionary Computation*, Piscataway, NJ, pp. 46–53, 2000.

[LAU 06] LAUMANNS M., THIELE L., ZITZLER E., "An efficient adaptive parameter variation scheme for metaheuristics based on the epsilon-constraint method", *European Journal of Operational Research*, vol. 169, pp. 932–942, 2006.

[LAW 77] LAWLER E., "Sequencing jobs to minimize total weighted completion time subject to precedence constraints", *Annals of Discrete Mathematics*, North-Holland, Amsterdam, vol. 2, pp. 343–362, 1977.

[LEM 07a] LEMESRE J., DHAENENS C., TALBI E., "An exact parallel method for a bi-objective permutation flowshop problem", *European Journal of Operational Research*, vol. 177, pp. 1641–1655, 2007.

[LEM 07b] LEMESRE J., DHAENENS C., TALBI E., "Parallel partitioning method (PPM): a new exact method to solve bi-objective problems", *Computers and Operations Research*, vol. 34, pp. 2450–2462, 2007.

[LEN 77] LENSTRA J., KAN A.R., BRUCKER P., "Complexity of machine scheduling problems", *Annals of Discrete Mathematics*, vol. 1, pp. 343–362, 1977.

[LEV 06] LEVITIN G., RUBINOVITZ J., SCHNITS B., "A genetic algorithm for robotic assembly line balancing", *European Journal of Operational Research*, vol. 168, pp. 811–825, 2006.

[LIN 97] LINDLER B., "Autosched tutorial", *Proceedings of the 29th Winter Simulation Conference*, Atlanta, GA, pp. 663–667, 1997.

[LIU 04] LIU M., WU C., "Genetic algorithm using sequence rule chain for multi-objective optimization in re-entrant micro-electronic production line", *Robotics and Computer Integrated Manufacturing*, vol. 20, pp. 225–236, 2004.

[LOD 04] LODI A., MARTELLO S., VIGO D., "Models and bounds for two-dimensional level packing problems", *Journal of Combinatorial Optimization*, vol. 8, pp. 363–379, 2004.

[LUT 98] LUTZ C., DAVIS K.R., SUN M., "Determining buffer location and size in production lines using tabu search", *European Journal of Operational Research*, vol. 106, pp. 301–316, 1998.

[LYO 97] LYONNET P., *La qualité: outils et méthodes*, Lavoisier, Paris, 1997.

[MAH 00] MAHDI H., Utilisation des métaheuristiques pour la résolution des problèmes d'agencement d'ateliers, de découpe 2D et d'ordonnancement, Thesis, School of Mines of Nancy, Nancy, 2000.

[MAK 05] MAKDESSIAN L., Structuration et choix d'équipements des lignes de production: approches mono et multicritère, Thesis, University of Technology of Troyes, Troyes, 2005.

[MAK 08a] MAKDESSIAN L., YALAOUI F., DOLGUI A., "Optimisation des lignes de production. Partie I: cas monocritère", *Journal of Decision Systems*, vol. 17, no. 3, pp. 313–336, 2008.

[MAK 08b] MAKDESSIAN L., YALAOUI F., DOLGUI A., "Optimisation des lignes de production. Partie II: une approche multicritère", *Journal of Decision Systems*, vol. 17, no. 3, pp. 337–368, 2008.

[MAR 80] MARTELLO S., TOTH P., *Knapsack Problems Algorithms and Computer Implementation*, John Wiley, New York, 1980.

[MAV 09] MAVROTAS G., "Effective implementation of the ε-constraint method in multi-objective mathematical programming problems", *Applied Mathematics and Computation*, vol. 213, pp. 455–465, 2009.

[MES 99] MESGHOUNI K., Application des algorithmes évolutionnistes dans les problèmes d'optimisation en ordonnancement de production, Thesis, University of Lille 1, Lille, 1999.

[NAH 06] NAHAS N., AIT-KADI D., NOURELFATH M., "A new approach for buffer allocation in unreliable production lines", *International Journal of Production Economics*, vol. 103, pp. 873–881, 2006.

[OSM 96] OSMAN I., LAPORTE G., "Metaheuristics: a bibliography", *Annals of Operations Research*, vol. 63, pp. 513–623, 1996.

[PAL 65] PALMER D., "Sequencing jobs through a multi-stage process in the minimum total time: a quick method of obtaining near optimum", *Operational Research Quarterly*, vol. 16, pp. 101–107, 1965.

[PAP 93] PAPADOPOULOS H., HEAVEY C., BROWNE J., *Queueing Theory in Manufacturing Systems Analysis and Design*, Chapman and Hall, London, 1993.

[PAP 01] PAPADOPOULOS H., VIDALIS M., "Minimizing WIP inventory in reliable production lines", *International Journal Production Economics*, vol. 70, pp. 185–197, 2001.

[PAR 96] PARETO V., *Cours d'économie politique*, F. Rouge, Lausanne, 1896.

[PEL 07] PELLEGRINI P., FAVARETTO D., MORETTI E., "Multiple ant colony optimization for a rich vehicle routing problem: a case study", *Knowledge-Based Intelligent Information and Engineering Systems*, Lecture Notes in Computer Science, vol. 4693, pp. 627–634, 2007.

[POR 96] PORTMANN M.-C., "Genetic algorithm and scheduling: a state of art and more propositions", *Proceedings of the Workshop on Production Planning and Control*, Fucal, Mons, Belgium, 1996.

[PRI 97] PRINS C., *Algorithmes de graphes*, Eyrolles, Paris, 1997.

[PRO 92] PROTH J., *Conception et gestion des systèmes de production*, Presses Universitaires de France, Paris, 1992.

[PRZ 06] PRZYBLSKI A., Méthode en deux phases pour la résolution exacte de problèmes d'optimisation combinatoire comportant plusieurs objectifs: nouveaux développements et application au problème d'affectation linéaire, Thesis, University of Nantes, Nantes, 2006.

[REE 93] REEVES C., *Modern Heuristic Techniques for Combinatorial Problems*, John Wiley, New York, 1993.

[REK 06] REKIEK B., DELCHAMBRE A., *Assembly Line Design*, Springer Verlag, London, 2006.

[RII 02] RIISE A., "Comparing genetic algorithms and tabu search for multi-objective optimization", *IFORS, Abstract Conference Proceedings*, 2002.

[RUB 91] RUBINOVITZ J., BUKCHIN J., "Design and balancing of robotic assembly lines", *Proceedings of the 4th World Conference Robotics Research*, Pittsburgh, PA, 1991.

[SAD 99] SADOWSKI D., BAPAT V., "The Arena product family: enterprise modeling solutions", *Proceedings of the 31th Winter Simulation Conference*, Phoenix, AZ, pp. 159–166, 1999.

[SCH 97] SCHOLL A., KLEIN R., "SALOME: a bidirectional branch and bound procedure for assembly line balancing", *INFORMS Journal of Computation*, vol. 9, no. 4, pp. 319–334, 1997.

[SOU 94] SOUILAH A., Les systèmes cellulaires de production : l'agencement inter-cellulaire, Thesis, University of Metz, Metz, 1994.

[SRI 94] SRINIVAS N., DEB K., "Multiobjective function optimization using non-dominated sorting genetic algorithms", *Evolutionary Computation*, vol. 2, no. 3, pp. 221–248, 1994.

[TAH 04] TAHAR D.N., CHU C., YALAOUI F., AMODEO L., "Ordonnancement dans un atelier d'impression de type job shop hybride", in *Proceedings 5e Conférence Francophone de Modélisation et Simulation. Modélisation et simulation pour l'analyse et l'optimisation des systèmes industriels et logistiques, MOSIM'04*, Nantes, France, pp. 755–762, 1–3 September, 2004.

[TAM 92] TAMAKI H., "Maintenance of diversity in a genetic algorithm and application to the job shop scheduling", *Proceedings of the IMACS/SICE Int. Symp. On MRP2*, pp. 869–869, 1992.

[ULU 95] ULUNGU E., TEGHEM J., "The two phase method: an efficient procedure to solve bi-objective combinatorial optimization problems", *Foundations of Computing and Decision Sciences*, vol. 20, no. 2, pp. 149–165, 1995.

[VAM 04] VAMANAN M., WANG Q., BATTA R., SZCZERBA R., "Integration of COTS software products arena and cplex for an inventory/logistic problem", *Computers and Operations Research*, vol. 31, pp. 533–547, 2004.

[WAG 58] WAGNER H., WHITIN T., "Dynamic version of the economic lot-size model", *Management Science*, vol. 5, pp. 89–96, 1958.

[WAS 07] WASCHER G., HAUSSNER H., SCHUMANN. H., "An improved typology of cutting and packing problems", *European Journal of Operational Research*, vol. 183, pp. 1109–1130, 2007.

[YAL 04] YALAOUI A., Allocation de fiabilité et de redondance dans les systèmes parallèle-série et série-parallèle, Thesis, University of Technology of Troyes, Troyes, 2004.

[YAL 06] YALAOUI F., CHU C., "A new exact method to solve the $Pm/r_i/\sum C_i$ problem", *International Journal of Production Economics*, vol. 100, no. 1, pp. 168–179, 2006.

[YAL 08] YALAOUI N., MAHDI H., AMODEO L., YALAOUI F., "Hybrid method for solving a layout problem", *Proceedings of the 8th International Conference on Computational Intelligence FLINS08*, Madrid, Spain, pp. 731–736, 2008.

[YAL 10] YALAOUI N., Agencement et ordonnancement d'un atelier de l'industrie automobile et aéronautique, Thesis, University of Technology of Troyes, Troyes, 2010.

[YAM 92] YAMADA T., NAKANO R., "A genetic algorithm applicable to large scale job shop problem", *Proceedings of the 2nd International Conference on Parallel Problem Solving from Nature*, Amsterdam, Netherlands, pp. 283–292, 1992.

[YAM 98] YAMASHITA H., ALTIOK T., "Buffer capacity allocation for a desired throughput in production lines", *IIE Transactions*, vol. 30, pp. 883–891, 1998.

[YAN 03] YANG T., KUO Y., CHOU P., "Optimization of physical flows in an automotive manufacturing plant: some experiments and issues", *Engineering Applications of Artificial Intelligence*, vol. 16, pp. 293–305, 2003.

[YAN 05] YANG T., KUO Y., CHOU P., "Solving a multiresponse simulation problem using a dual-response system and scatter search method", *Simulation Modelling Practice and Theory*, vol. 13, no. 4, pp. 356–369, 2005.

[ZHA 04] ZHANG H., LI H., "Simulation-based optimization for dynamic resource allocation", *Automation In Construction*, vol. 13, pp. 409–420, 2004.

[ZIT 98] ZITZLER E., THIELE L., An evolutionary algorithm for multiobjective optimization: the strength pareto approach, Technical Report 43, Zurich, Switzerland, 1998.

[ZIT 99] ZITZLER E., THIELE L., "Multi-objective evolutionary algorithms: a comparative study and strength Pareto approach", *IEEE Transaction on Evolutionary Computation*, vol. 3, no. 2, pp. 257–271, 1999.

[ZIT 01] ZITZLER E., LAUMANNS M., THIELE L., SPEA2: improving the strength pareto evolutionary algorithm, Report 103, Computer Engineering and Networks Laboratory (TIK), ETH Zurich, Switzerland, 2001.

Index